Acclaim for

Marie Winn's

Red-Tails in Love

"Engaging . . . Dr. Zhivago with feathers. . . . That such simple pleasures can be savored today, in the heart of frantic New York City, is a bit of a miracle." —*The Boston Globe*

"'Give your heart to the hawks,' the poet Robinson Jeffers wrote. Marie Winn certainly has, and so will readers of this delightful book." —*The New York Times Book Review*

"A delightful account of how nature flourishes in the most unlikely of places. . . . [Winn's] infectious enthusiasm for all things feathered makes every warbler, owl, or starling dear." —*Elle*

"Engaging and exciting. . . . If it seems difficult to imagine much drama in the daily rounds of a birdwatcher, well, you haven't yet read this book." —*BookPage*

"Enchanting." —*The New York Review of Books*

"This book has a rare charm, beguiling the innocent reader who thinks it's about birdwatching into a wonderland of many levels." —*Robert MacNeil*

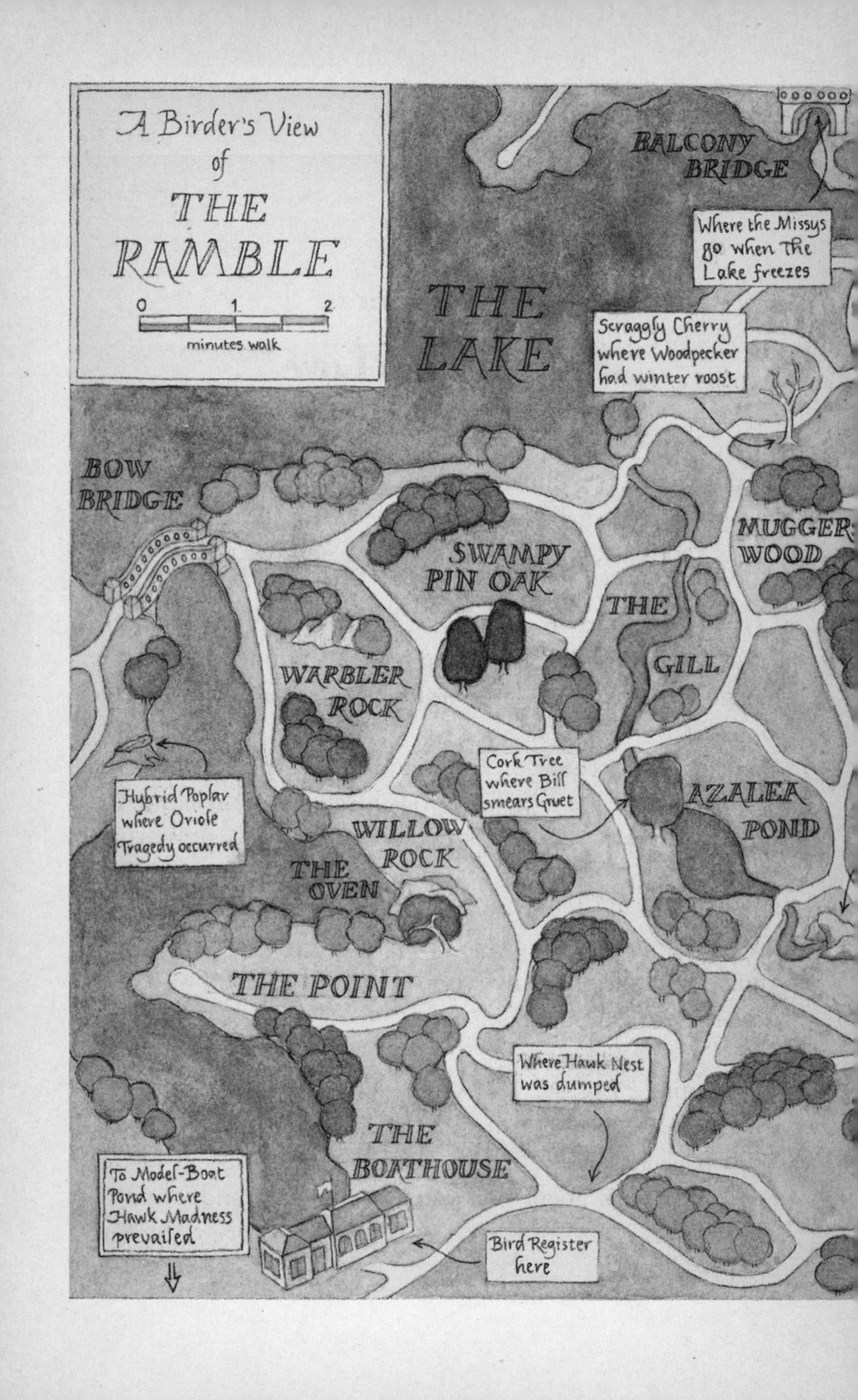
A Birder's View of THE RAMBLE
0 1 2
minutes walk
THE LAKE
BALCONY BRIDGE
Where the Missys go when The Lake freezes
Scraggly Cherry where Woodpecker had winter roost
BOW BRIDGE
SWAMPY PIN OAK
MUGGER'S WOOD
THE GILL
WARBLER ROCK
Cork Tree where Bill smears Gruet
AZALEA POND
Hybrid Poplar where Oriole Tragedy occurred
WILLOW ROCK
THE OVEN
THE POINT
Where Hawk Nest was dumped
THE BOATHOUSE
To Model-Boat Pond where Hawk Madness prevailed
Bird Register here

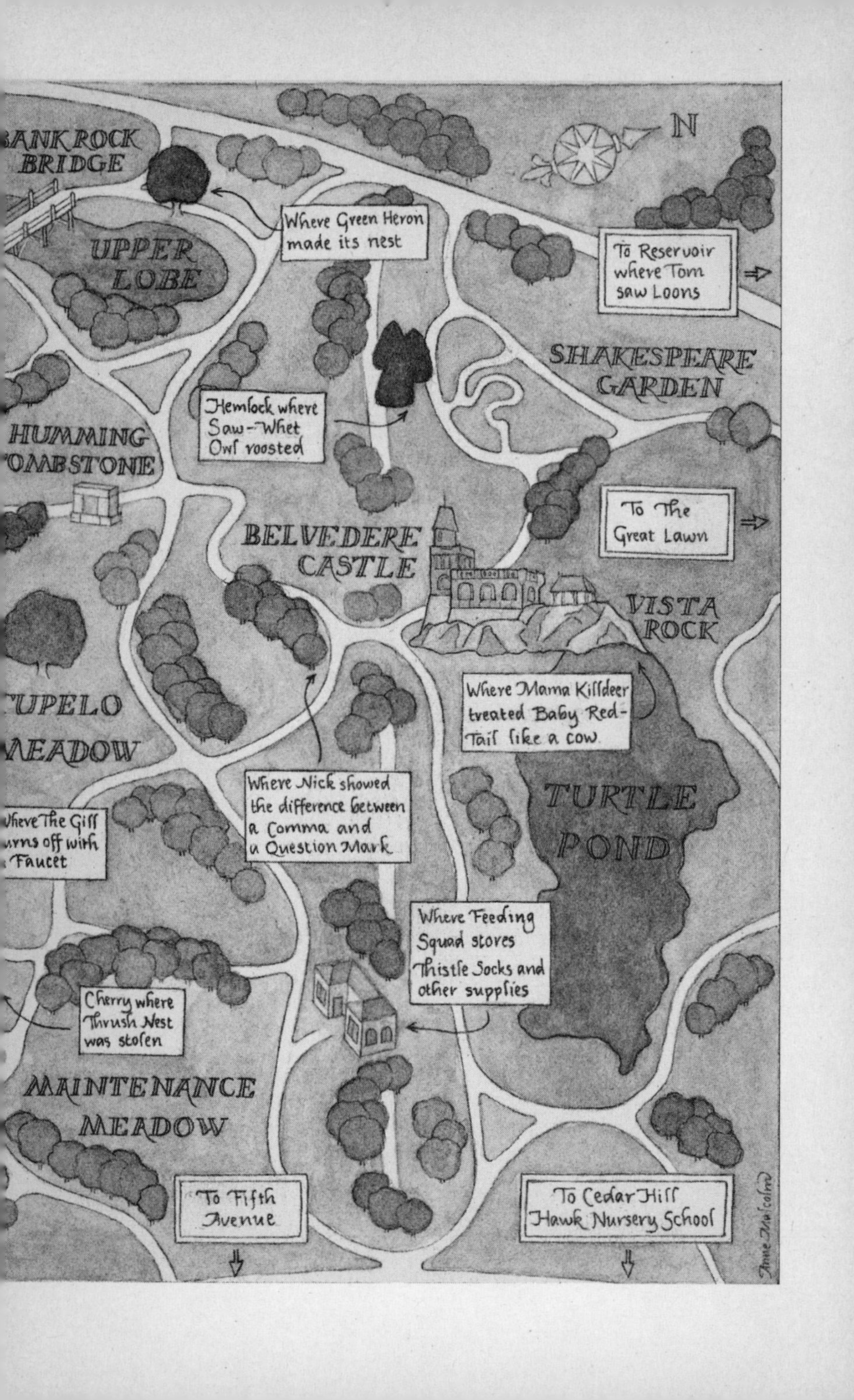
BRIDGE
UPPER LOBE
Where Green Heron made its nest
N
To Reservoir where Tom saw Loons
SHAKESPEARE GARDEN
Hemlock where Saw-Whet Owl roosted
BELVEDERE CASTLE
To The Great Lawn
VISTA ROCK
Where Mama Killdeer treated Baby Red-Tail like a cow
TURTLE POND
Where Nick showed the difference between a Comma and a Question Mark
Where Feeding Squad stores Thistle Socks and other supplies
Cherry where Thrush Nest was stolen
MAINTENANCE MEADOW
To Fifth Avenue
To Cedar Hill Hawk Nursery School

Also by MARIE WINN

The Plug-In Drug: Television, Children and the Family

Unplugging the Plug-In Drug

Children Without Childhood

The Baby Reader

Marie Winn

Red-Tails in Love

Marie Winn writes a column on nature and bird-watching for *The Wall Street Journal.* Among her twelve previous books are *The Plug-In Drug: Television, Children and the Family,* and *Children Without Childhood.* Married to the filmmaker and palindromist Allan Miller, she spends part of every day in Central Park.

Red-Tails in Love

A Wildlife Drama

IN

Central Park

Marie Winn

VINTAGE DEPARTURES

VINTAGE BOOKS

A DIVISION OF RANDOM HOUSE, INC. / NEW YORK

FIRST VINTAGE DEPARTURES EDITION, APRIL 1999

 Published in the United States by Vintage Books, a division of Random House, Inc., New York, and simultaneously in Canada by Random House of Canada Limited, Toronto. Originally published in hardcover in the United States by Pantheon Books, a division of Random House, Inc., New York, in 1998.

Portions of this work were originally published in somewhat different form in *Smithsonian* magazine and *The Wall Street Journal.*

The Library of Congress has cataloged the Pantheon edition as follows:
Winn, Marie.
Red-tails in love : a wildlife drama in Central Park / Marie Winn.
p. cm.
Includes index.
ISBN 0-679-43997-8
1. Bird watching—New York (State)—New York.
2. Red-tailed hawk—New York (State)—New York.
3. Central Park (New York, N.Y.)
I. Title.
QL684.N7W55 1998
598'.07'2347471—dc21 97-28417
CIP

Vintage ISBN: 0-679-75846-1

Author photograph © Charles Kennedy

Book design by M. Kristen Bearse

Endpapers map by Anne Malcolm
Map copyright © 1998 by David Lindroth, Inc.

www.randomhouse.com/vintage

Printed in the United States of America
10 9 8 7 6 5

for Joe

Our own absence, the only certain thing
before we came into this world, or after our death.
Hence the pleasure of recognizing the infinite variety
of what is other than us . . .

ITALO CALVINO

Contents

A Note about Nomenclature

❧

Standard dictionaries spell the names of all birds from auks to whip-poor-wills with lower-case letters, while the American Ornithologists' Union, an authority scientists and serious birdwatchers follow, requires that bird names be capitalized.

There are good reasons behind the AOU's policy. Many bird names include words that describe salient features of that species—the yellow warbler, for example. When the name is spelled in lower-case letters, the reader cannot be sure whether it refers to a specific warbler of the species *Dendroica petechia* or an unidentified little bird that happens to be yellow. Calling it a Yellow Warbler solves the problem.

Yet it's hard to read a story when each page is interrupted by great numbers of capitalized words. For this reason (with apologies to the AOU and to punctilious birdwatchers), I decided to go with the dictionary. Whenever confusion threatens as a result of lower-case spelling (as in the story of the gray gull on page 174), I throw in the bird's taxonomic designation to make my meaning clear.

Red-Tails in Love

This book tells the story of six years in Central Park I spent with a faithful band of birdwatchers and nature lovers—the Regulars. It is about the park and its wildlife—and a pair of hawks that captured our hearts.

I say the story went on for six years, because that's when I stopped to write it all down. But the story hadn't really ended then, only one exciting part of it. It's still going on as I write and may go on forever. You can go to Central Park and see for yourself.

PROLOGUE

Falling in Love

Loving helps us to discern, to discriminate. The bird-lover in a wood at once distinguishes the twittering of different species, which to ordinary people sound the same.

MARCEL PROUST

SCENE ONE

The Bird Register

❧

If it is possible to fall in love with a thing, I believe I fell in love with the Bird Register the day I first opened it. The emotions were familiar: the same feeling of excitement, of undeserved luck, the mildly deluded sensation that a new kind of happiness was just around the corner, the certainty that life was about to divide forever into a before and after.

The Loeb Boathouse, a nondescript building located at the east end of Central Park's rowboat lake, is where the Register resides, though not always in the same place. During the years I've known it, the Bird Book, as it is often called, has lived on the frozen-yogurt bar, on a shelf behind it, and on the cafeteria counter where the little packets of sugar, mayonnaise, mustard, and grape jelly are kept. Currently it may be found behind the podium where reservations are taken for the Boathouse Café, a private restaurant. It may have moved again by the time you read this, but keep looking. It's sure to be there, somewhere, sitting right out in the open as if it were an inconsequential thing instead of a local tribe's central treasure.

I remember casually picking up the plain, blue canvas loose-leaf notebook with its sloppily hand-lettered legend on the front cover: CENTRAL PARK BIRD REGISTER AND NATURE NOTES: ENJOY BUT PLEASE DO NOT REMOVE. I opened it for a quick glance at its contents. Then, with that greedy feeling one gets after cautiously tasting some unpromising new dish and discovering it to be delicious, I stood there devouring page after page.

I had known there were robins and sparrows and blue jays in Central Park. I had even seen a warbler or two on occasion. Now I read of owls and snipes, goshawks and scarlet tanagers, flycatchers, vireos, kinglets, and twenty, thirty species of warbler—all, it appeared, more accessible than in any wild forest or meadow.

Squirrels, rats, and dogs were the only mammals I had encountered in my past visits to the park. Here were raccoons and woodchucks and bats. And snapping turtles laying eggs. And bullfrogs croaking at dawn. And butterflies and dragonflies. And so much more, all to be found at such intriguing locations as the Humming Tombstone, Willow Rock, the Oven, Muggers Woods, the Point, the Azalea Pond. Where were these places? I wanted to find them. They weren't on any Central Park map.

The detailed observations, notations, exhortations, invitations, descriptions, maps, diagrams, even poems in the Bird Register gave me a tantalizing glimpse not only of the unexpected wildlife treasures of Central Park but of a community as well. Who were these people? I longed to know them, to learn their secrets. And there was the Bird Register right out in the open. "Don't be an eavesdropper," its voices seemed to be saying. "Come and join us, come and learn."

SCENE TWO

Into the Woods

Everyone in the birdwatching tribe knows Sarah Elliott. A trim, redheaded, trenchant woman in her sixties, Sarah once roamed the park with a different band of Regulars from those active today. She remains a vital link with the past, for it was she who started the Bird Register in the first place.

A native of Chicago, Sarah was not solidly hooked on birds until she moved to New York in the early 1960s. There she began birdwatching in earnest, learning to identify birds in the company of some of the city's top birders of the time: Richard Harrison, Dick Plunkett, Bert Hale. Central Park was where most of her bird studies took place. There, in 1972, she met Lambert Pohner and began her journey from birdwatcher to naturalist—a person who studies nature in all its forms.

Lambert Pohner's obituary in *The New York Times* on July 13, 1986, described him as "an elf of a man, with a white beard and a bush hat . . . who watched over the birds and butterflies of Central Park for more than 40 years." Sarah had often wandered through the park with Lambert, picking up an approach to learning that appealed to her, one that took in the whole picture—the trees, flowers, frogs, turtles, butterflies of the park, as well as the birds.

Sarah soon revealed an organizing skill all her own. As she rambled through the woodlands, she kept two lists: one, of the birds she had seen that day; the other, of the birdwatchers she ran into. Until then, Central Park's birders knew each other slightly, or not at all. Sarah became a common link. Around 1975 Sarah took the crucial step that marked the birdwatching community's true beginning: She started the Bird Register. Now Central Park's birdwatchers had a place to meet, if only on paper.

In 1980, the Central Park Conservancy, then a newly formed organization seeking to rehabilitate a down-at-the-heels park, asked Lambert to lead bird walks in the Ramble, a 37-acre wilderness in the heart of the park. He accepted the invitation, and invited Sarah to share the job.

There had been regular birdwatching walks in the park before Lambert and Sarah's. There was the legendary Farida A. Wiley, who began leading walks in 1938 under the aegis of the American Museum of Natural History and continued for almost fifty years. Lambert and Sarah's walks were different, slower, as likely to focus on a plant or grasshopper or bat or raccoon as on a bird.

After Lambert's death, Sarah kept his memory alive by continuing the Wednesday and Sunday morning walks during the spring and fall migrations, on her own now, but still in Lambert's uncompetitive, reflective style. Many of the park's most ardent birdwatchers and nature lovers first caught the Central Park bug, as it were, in one of Sarah Elliott's bird classes.

THOUGH SARAH WAS to be my entry into the vibrant world of Central Park's birdwatchers, it was dead birds that first brought us together. That was in late May of 1991. I had come upon an article she'd written in the New York City Audubon

Society newsletter in which she'd declared that the city's brightly illuminated skyscrapers were deathtraps for passing migratory birds. To save birds' lives, she proposed a letter-writing campaign: Write to the owners of well-lit skyscrapers to tell them to turn off their bright lights during the migration seasons, at least on foggy and rainy nights, she instructed.

I had begun writing an occasional column about birds for *The Wall Street Journal*. I called her for an interview about her crusade and she chuckled at the opportunity to reach a nice pack of fat-cat skyscraper owners.

We arranged to go on a fact-finding expedition the next foggy or rainy day. Our mission: to look for dead birds near the Empire State Building, the Met Life Building, the World Trade Center, and other illuminated towers.

On a highly propitious (i.e., dismal) morning a few days later our inspection tour turned up no dead birds at all, merely one dazed warbler that zipped off the moment we tried to pick it up. Perhaps Sarah thought she owed me more birds, for as we parted that morning she offered to take me on a bird walk in Central Park some day. I called her that very afternoon to set a definite date. Open Sesame.

❧

WE MET at the Boathouse. There's an expensive restaurant at one end where tourists and swanky New Yorkers congregate. The Regulars, a little band of the park's most devoted birdwatchers and nature lovers, prefer the plain cafeteria next door. That's where they warm up on cold days, take shelter from rain, and find out where the action is. The vegetable soup isn't bad, the blueberry muffins are homemade, and the Bird Register is kept there—the major attraction.

From the moment we set off, I began scribbling notes and drawing maps filled with kindergarten-style representations of rocks, fences, and lampposts where a path to somewhere or other is to be found. It was my first walk in the Ramble and I wanted to remember what I was learning—the names of trees, flowers, birds, streams, bridges, people, everything!

Most of all I wanted to be sure I wouldn't get lost when I came the next time. This was Central Park, after all, and everyone knows it's not the safest place in the world. I'd been to the park many times before, to be sure, for I grew up nearby, but I had never been in the Ramble. That dense woodland had always been out-of-bounds—a scary place. The feeling still lingered on that day in mid-June.

As Sarah Elliott led me up the steep footpath that begins behind the Boathouse at lamppost 7401 and leads into the Ramble, I learned the first of many park secrets: the first two digits on each lamppost tell its location relative to the nearest city street. (North of 99th Street the numbers begin with 00, indicating 100th Street.) In this case it revealed that we were somewhere near 74th Street. For the park's birdwatchers the numbers serve to pinpoint important spots where birds have been sighted so others can easily find them.

We were hardly halfway to the top when the show began. Suddenly a pair of tufted titmice appeared on a branch just ahead and seemed to be keeping up with us as we walked. They were pronouncing their raucous version of the black-capped chickadee call as they flew from branch to branch. Dee-dee-dee!

As Sarah stopped and fumbled with something in her bag, the titmice hopped to a branch so near I could no longer focus my binoculars on them. The birds continued yelling, making little forays out towards us and then back onto the branch.

Sarah unscrewed a small black plastic film canister, removed a peanut fragment, and held it out on her hand. One of the titmice promptly landed and snatched the peanut away. She provided another peanut tidbit for me to do the same. I'm embarrassed to find in my notes that a bird's feet on the palm of the hand feel "like fairy wings." In years to come I was to see this little drama many times, for most of the Regulars hand-feed the resident birds on occasion. Chickadees, blue jays, and cardinals are others that yammer to be fed when the Regulars walk by, though they only come close; titmice and chickadees alone actually come to the outstretched hand, with a downy woodpecker taking the plunge once in a blue moon.

The Ramble proper starts at the top of the hill and an unmistakable landmark—the Balancing Boulder—marks its beginning. One of the park's many naturalistic artifices, it is a huge upended rock that from a certain angle appears to be balancing so precariously on top of another, horizontal boulder that a good, firm push should send it toppling. Children seem compelled to give it a try.

Just past the boulder, at a crossroads of sorts where three paths diverge, we bore left. "You'll see a few warblers in a moment," Sarah promised. And so we did, soon after we arrived at the Point, a wooded promontory jutting out into the rowboat lake. This little spit of land is a famous birdwatching spot during the spring and fall migrations.

Standing at an elevation high above a heavily wooded depression just to its west, known as the Oven, observers at the Point can look down into the crowns of the Oven's oaks and willows and get an exceptionally good look at such elusive warblers as the Cape May and the cerulean, treetop feeders. Thus birdwatchers can avoid the occupational hazard known as warbler neck, a painful condition brought on by long-term upward gazing.

Birdwatchers have long joked about warbler neck, but the

ailment may be more serious than people believe. Neurologists have discovered that when people assume extreme neck positions for extended periods of time, blood flow through the vertebral arteries is reduced, leading to an increased vulnerability to strokes. Researchers singled out the tilted-back head position of women having a shampoo at a beauty salon. But it happens to be the very posture of birdwatchers craning to see a warbler at the top of a tree.

At the Point, according to my notes: "2 magnolia warblers, female redstart, blackpoll warbler; warblers eat berries in fall, otherwise bugs; big uproar about Point and Ramble in '81 when Conservancy chopped down trees to restore historic views—birdwatchers still mad."

Birders sometimes see more than twenty species of warblers at the Point, Sarah told me. But by the beginning of June the spring migration is winding down. The birds we saw that day were the stragglers—mostly first-year birds (last summer's fledglings) and females. The bright-colored males in breeding plumage had raced ahead to their breeding grounds farther north to get the choicest nesting spots, Sarah said. The females would join them a week or so later, after the males had worked out their territorial disputes.

From the Point, we walked a short distance to Willow Rock, a flat outcrop high above a peaceful lobe of the Lake. It was named for the two thick black willows growing nearby, Sarah explained. (One of them fell in a storm in 1994.) There was a small tree, almost a sapling, growing out of a crevice in the rock, near the edge. "Look at this," said Sarah. "It's a real peach tree. Maybe somebody planted a pit here." I could see dozens of unmistakable mini-peaches growing all over the little tree. They were still green—though already covered with that characteristic fuzzy down botanists call pubescence.

Willow Rock offered easy views of the *same* treetop activity we had seen from the Point but from a different angle, as Willow Rock is directly across from it. What a splendid opportunity this affords birdwatchers: with the sun behind them they can spend their morning hours gazing at warblers from the Point, and then do the same at Willow Rock in the afternoons.

Like many impassioned novices, I wanted to know everything all at once—the names of every plant, every bird, every *part* of a bird. Sarah, it was clear, did not suffer such acquisitive fools gladly. "What's that tree with the bunches of red berries?" I asked—it was probably my hundredth question. "Oh, that's the bunches-of-red-berries tree," she answered with a smile. A few moments later, as we saw a black-crowned night heron land on a willow overhanging the Lake, I wondered out loud about one of the bird's most conspicuous features: "Is that called a bill or a beak?" I asked Sarah. "Yes," she answered firmly, and that was that.

Her message sank in. Don't worry about nomenclature when a bird is sitting in front of your nose. Look at it. Notice everything you can about it—its yellow-green legs, its blackish back and cap. Look at that orange eye. Later I checked out the bill vs. beak question in *The Birdwatcher's Companion,* a reliable reference book by Christopher Leahy. His entry for *beak* says: "Essentially synonymous with *bill.* In more restrictive usage, refers particularly to larger bills, especially the hooked beaks (or bills) of birds of prey. In general 'bill' is the preferred term in ornithological/birdwatching contexts." Yes.

Next we walked up a small incline, passing another large boulder—Warbler Rock, Sarah called it. I was growing more and more uncertain of our orientation as we veered away from the Lake and open sky and entered the deeper woods. Soon the leafy canopy was closed all around us. Even the air felt different—more carbon dioxide, I imagined.

Once in the woods, I gave up mapmaking entirely. I had

completely lost my sense of direction and was too embarrassed to keep asking Sarah which way was which. In other parts of Central Park one can orient oneself by the surrounding tall buildings—the Fifth Avenue skyline is east, the twin towers of the El Dorado and the San Remo are west. But in the heart of the Ramble the city has vanished; all reminders of civilization are obscured by trees.

An illusion begins to take over: You are in an enchanted woodland. Even the park furniture seems to belong. "Sit on us," the dark green benches command. "Look at the birds, look at that flower. Stay awhile. Don't hurry or you'll miss something." The black cast-iron lampposts no longer foreshadow the feared nightmares of Central Park in the dark. At nightfall their soft light will show owls and bats and gaudy moths attracted as if by moonlight.

We had entered a virtual maze of little paths, all unmarked, winding, twisting, taking us past ravines, waterfalls, rustic benches, scenic vistas. By then my notes had become as confusing as the Ramble itself, a jumble of bird names, people names, plant names, and samples of Sarah's botany and ornithology sound bites: "parula, magnolia, Wilson's—warblers, Ruth, George, Ira, Dave—Regulars, double-file viburnum has 2 rows of white flowers in May, Swainson's thrush has buffy eye-ring, gray-cheeked thrush doesn't, spice bush—smells good, sassafras has 3 kinds of leaves, grackles walk, crows hop."

We reached the area birdwatchers call the Swampy Pin Oak (sticklers prefer to call it the Pin Oak swamp, for there is no such tree as a swampy pin oak). Within a little grove of trees growing in a moist sumpy spot, there is one significantly larger tree right in the center, *the* "swampy" pin oak. That day it was hopping with a variety of small birds while the wet ground below revealed others busily poking around in the mud. Watching the action from two ringside benches were a variety of birdwatchers as well: Alice and Ira, Max and Nellie, David Monk, Sheila and Lou, Mary Birchard, Chris and Marianne, Judy—some of them, I soon learned, were Regulars, others Seasonal Migrants, birdwatchers who show up regularly only during the migration seasons.

There I learned yet another secret from Sarah. The eggs of minute aphids hatch on the leaves of pin oaks in May and June, attracting a great variety of migrating songbirds. So head for the pin oaks if you want to score heavily during spring migration.

On our way to the Azalea Pond—our last destination, said Sarah—we came to a couple of thick holly trees just to the right of the path. M. M. Graff, author of *Tree Trails in Central Park,* disapproved of these particular trees, accusing them of being "a gloomy black-green at best and made even more funereal by a coating of city soot." They looked shiny and handsome to me. I was even more taken with them when Sarah told me these were a favorite roosting place for owls. I kept returning to the spot over the next few weeks, hoping to find a roosting owl, until a kindly birdwatcher informed me that owls show up in Central Park only in the late fall and winter.

After turning right at the second holly tree, we found ourselves out of the deepest woods. Now the skyline of Central Park West was visible once again. Walking northward along the same path, we soon came to a grassy knoll where a small group of birdwatchers were standing in a classic pose: looking upward, watching something invisible to the naked eye. Their binoculars all pointed to the same spot in the same tangle of wisteria vines. Sarah did not ask, "What are you looking at?" as I might have done, but maintained birders' etiquette, merely raising her own binoculars in the same direction. It took her hardly a minute to locate the bird—a "good bird," she told me with some excitement. "A hooded warbler." A good bird? All the other birds we had seen on our walk had been just as good, as far as I was concerned.

At the edge of that little clearing was a large, rectangular granite block. "Do you hear anything coming from that block?" Sarah asked me. At first I heard nothing but the sounds of birds all around and dogs barking in the distance. Then I did hear it—a faint buzzing sound. "We call that the Humming Tombstone," Sarah said, and I immediately realized that the block did resemble a large cemetery monument. "Some birders use the sound as a hearing test. Every year they see how near they have to get before they hear the buzz."

I stepped away until I couldn't hear it, and noted the place. I still check every year to see if I hear the buzz at the same spot. So far so good.

It took me years to discover what makes the Humming Tombstone hum: a mechanism within that controls all the lights in the vicinity, turning them off in the morning and on at night. One year in the early spring I found myself at the Tombstone at sunrise. I thought I had gone deaf, for it was silent. A few moments later, at 6:00 a.m. sharp, it started to buzz.

After another few twists and turns of the path where we encountered, according to my notes, "common yellowthroat, ovenbird, three raccoon babies up a tree, Mo & Sylvia," we crossed a small rustic bridge known as (surprise) the Rustic Bridge and I began to hear a crescendo of bird sounds. We were approaching the Azalea Pond feeding station.

The Azalea Pond is a small body of water fed by the Gill, one of the Ramble's most picturesque features. This replica of a meandering stream is turned on and off by a hidden faucet. Though winter was long over, and official bird-feeding would not resume until cold weather set in, the place was still well stocked: birdseed was scattered all around, bits of bread, and some brownish lumps that looked like dry dog food. (They were.) Chunks of suet had been attached to nearby tree trunks. And birds were everywhere.

Sparrows, pigeons, cowbirds, and mourning doves were eating seeds on the bare ground in front of six scraggly bushes. These were the azaleas for which the place was named, a rather garish carmine variety called "hinode-giri," which M. M. Graff called "an offense to the eye in almost any garden setting and a shrieking dissonance in this quiet spot."

As I took in the scene, entranced, other birds flew in and out, mainly titmice and cardinals, grabbing bits of peanuts from a sardine tin attached to the trunk of an oak. Several woodpeckers were working the suet, squirrels were racing up and down the tree trunks, trying to get at the peanuts, while a man with a shock of white hair threw sticks at them and shouted, "Go away!" After the bucolic serenity of Willow Rock and the Swampy Pin Oak, the place seemed a madhouse of activity.

Birdwatchers were everywhere too. The Azalea Pond is one of their major gathering spots, and some we had met earlier were now sitting on the benches facing the feeders. "George, Joe Richner (keeps list of birdwatchers, not birds)—white-breasted nuthatch, red-bellied woodpecker goes Chork!" read my notes.

Sarah was on her way home, but I decided to stay just a few more minutes. "I'll find my way out," I assured her, knowing it wouldn't be easy. Bill, the white-haired man who had been throwing sticks at the squirrels, was now throwing peanuts to a pair of cardinals. "Get this!" he was saying with each throw. The male was bolder and got more than the female. Joe Richner put my name on his People list. "I have four Marys and you're the second Marie," he informed me. A red-bellied woodpecker arrived and grabbed one of the cardinal's peanuts. George and Ira were talking about vitamins. Before I knew it, an hour had gone by and I was still at the Azalea Pond. Hours pass like minutes there.

A spell must have come over me in that dark, mysterious wood, for I came back the next day and the next, and never stopped coming. I still get lost in the Ramble at times, and most of the Regulars admit that they do too.

SCENE THREE

From Fake to Real

❧

As Central Park was being planned in the middle of the nineteenth century, neither Frederick Law Olmsted, its principal designer, nor Calvert Vaux, his architect partner, intended any part of it to be a *real* wilderness. Wilderness was not a romanticized ideal in that era. Too recently had citizens of the young Republic struggled to survive in the untrammeled wilds of the New World to regard it as anything other than an obstacle to life and happiness. Besides, with plenty of unspoiled woodlands, meadows, and marshes an easy carriage drive away, why bring the mess right into one's front yard? Central Park was created as an improvement on the wild, a carefully fashioned landscape where city dwellers could come and enjoy the illusion of wilderness without any of its inconveniences or dangers.

A description of the Ramble written in 1878 gives an idea of the kind of stage-set wilderness the park provided in its infancy:

> In sauntering through the Ramble one comes upon bits of open, sunny lawn where perchance a gorgeous peacock is grandly trailing his long tail feathers over the short, soft grass. . . . Strange notes are heard from the thicket; there are guinea

> fowl, white turkeys, pigeons and other varieties of feathered chatterers which make their home there . . . lively chipmunks spring about from tree to tree. . . . Little streamlets flow under bridges covered with hanging vines and tumble in tiny cascades downward toward the Lake. In the season the thickets are brilliant with rhododendron.

Today the rhododendrons are few and far between. Instead, Japanese knotweed, one of Vaux and Olmsted's most ill-considered plantings, has taken over their carefully planned landscape. Yet this hardy, invasive non-native plant may be the only one to withstand the trampling of millions of visitors each year. The Arcadian bowers have been taken over by the homeless. The tame peacocks and guinea hens are long gone, their care and feeding no longer a part of the park's budget.

All the while, other creatures have been infiltrating those man-made woods and manicured lawns—raccoons, woodchucks, rabbits, frogs, turtles, butterflies, dragonflies, crickets, and birds—great numbers of birds. Gradually, through Nature's mysterious alchemy, the former fake has begun to turn into the real thing.

Who knows how the raccoons and woodchucks made their way to Central Park. Maybe some of them were abandoned pets. Others may have traveled down from the suburbs, crossing city streets in the dead of night, wandering hither and yon until they found themselves in . . . Shangri-La! And the birds? As the real country shrank and disappeared, as wilderness became a rare commodity, Central Park grew ever more alluring—a green oasis in the concrete desert.

The park's first official bird census, undertaken in 1886, listed 121 species. By 1996 the total had more than doubled—275 bird species on the list of *Birds of Central Park* published that year. Though the century has seen a northward expansion of certain bird species—tufted titmice, cardinals, mockingbirds, red-bellied woodpeckers were once birds that never appeared as far north as New York City—this cannot account for so great an increase. Habitat change in the park together with habitat loss in the environs has played a significant role.

The Ramble's transformation did not occur by plan. Neglect created the unmown meadows that soon became sparrow heavens, the unpruned trees and dead snags offering prime real estate for woodpeckers and raccoons—neglect dictated by New York City's ever-increasing budget constraints.

As the metamorphosis of fake to real progressed, word began to spread among bird lovers: Central Park was a great birdwatching area. Great? Yes. In a recent list for birders compiled by the bird expert Roger F. Pasquier, New York's 843-acre enclave of green was designated one of America's fourteen great birdwatching locales, together with Yosemite, the Everglades, Cape May, and other famed birders' meccas.

To be sure, not everyone was enchanted by the transformation. What is wildness to some seems a mess to others. That great defender of Vaux and Olmsted's vision, none other than the strongminded M. M. Graff, has agitated for years to restore the Ramble to its original design concept. In *Tree Trails,* she calls the Ramble "an eyesore, an overgrown weedy jungle with a few rotting trees to recall its intended character. Civilized people stay away; inevitably, the criminal element moves in." Clearly Ms. Graff does not include birdwatchers among the ranks of the civilized, though the binocular band easily outnumbers the criminals in Central Park—at least on a fine day in May.

In the past Graff's strictly horticultural attitude was shared by the Central Park Conservancy. In 1981 the Conservancy initiated a costly project in the Ramble intended to reclaim views of Bethesda Fountain and Belvedere Castle established in Vaux and Olmsted's original design. Unfortunately this required the felling of dozens of healthy, mature, bird-friendly trees that had grown bigger, as trees do, in the course of a century. Among them were a seventy-year-old chestnut oak, a cherry tree that flowered in two different colors, a persimmon, a magnolia that had once housed a family of cardinals, some hawthorns, locusts, and many black cherries. The park's bird lovers were appalled—they *knew* those trees personally! They protested effectively, capturing the attention of the news media and creating a public-relations nightmare for the Conservancy. The project was abandoned. Today, in a new ecology-minded era, the Conservancy's interest in the woodlands as a wildlife habitat is second only to the birdwatchers' own.

WHAT, IN FACT, accounts for Central Park's exalted position in Roger Pasquier's birdwatching pantheon? During the spring and fall migrations, when millions of songbirds take to the air, they need stopover places on the way for rest and refueling. Since most migrants travel at night, those that find themselves flying over Manhattan just before dawn don't have much choice: They funnel into Central Park in huge numbers. (Other city parks get their fill of migrants as well.)

This concentration of a great variety of birds in a small geographic area, a migrant trap, as it is called, leads to exceptional birdwatching. At the height of a recent spring migration, some of the park's best birdwatchers came up with a total of 124 species of birds in Central Park, the highest number of sightings in a single day up to then. The number may be higher by now.

During the spring or fall migrations, the wooded areas of the park are filled with birders as well as birds. Walking through the Ramble on a nice day in May or September, you might run into several members of the ornithology department at the American Museum of Natural History, for instance. They like to spend their lunch hours warbler-watching, a birdman's holiday, you could call it.

This concentration of expert birders in a single bird-rich location leads to a phenomenon called the Patagonia Picnic Table Effect, named after a rest stop in Patagonia, Arizona, where a rare bird was once discovered. Much to the amazement of the birders who rushed to the spot, three other equally rare birds showed up at the same roadside rest area on the same day. Was this a bizarre coincidence? A better explanation has become part of birdwatching mythology. Since so many experts had congregated at a single spot to look at the first rarity, their sharp eyes and ears spotted other unusual birds lurking in the vicinity.

In Central Park, similarly, the presence of so many accomplished birdwatchers helps to uncover elusive creatures that might otherwise go undiscovered. A hard-to-locate bird like the Connecticut warbler, considered a rare species in other locales, is regularly found in Central Park.

And of course you will inevitably run into some of the Regulars. A friendly lot, they're likely to take you in hand and steer you to where the action is. "There's a nighthawk sleeping on a horizontal branch just a little ways from here," they'll tell you if you give them half a chance, and even lead you to the bird. Many other places have abundant bird populations. But few make their birds so accessible.

Yet the greatness of Central Park has another, deeper source: the very idea that wildlife can exist and even thrive in the middle of a city like New York. It seems remarkable that a pair of wood thrushes, a diminishing species in America, birds of deep woods and sylvan glades, would choose to build a nest and raise a family in Central Park's Ramble, as they did a few years ago. On early May and June mornings and then just around dusk, the Ramble wood thrush gave regular concerts. And numbers of city people passing through stopped and listened to its penetrating, flutelike, heart-stoppingly beautiful song: Ee-oh-lee, ee-oh-loo-ee-lee, ee-lay-loo.

On June 22, 1853, before Central Park even existed on paper, Henry David Thoreau heard a wood thrush singing as he took his evening walk on Fair Haven Hill near Walden. He described the experience in his *Journal:*

> All that is ripest and fairest in the wilderness is preserved and transmitted to us in the strain of the wood thrush. This is the only bird whose note affects me like music, affects the flow and tenor of my thought, my fancy and imagination. It lifts and exhilarates me. It is inspiring. It is a medicative draught to my soul.

Though the wood thrush in the Ramble sings its song in the heart of the city, its mysterious power to evoke deep woods and the wildness of nature is undiminished. As Thoreau explained it (*Journal,* August 30, 1856): "It is vain to dream of a wildness distant from ourselves. There is none such. It is the bog in our brain and bowel, the primitive vigor of Nature in us, that inspires that dream."

SCENE FOUR

The Regulars

Just as the physical realities of Central Park—its plants, trees, lakes, streams, and the wildlife within—stand in sharp contrast to the man-made world around it, so the park's little band of Regulars stand in relief to the culture of urban life.

The passage of time is different for the Regulars. The changing length of days, the arrivals and departures of birds, the flowering of plants, and the changing colors of leaves organize their time far more than the calendars and clocks and schedules of contemporary life.

The seasons do not begin for them on dates given in almanacs. Spring begins with the arrival of the woodcock in February, or at the moment the first phoebe begins hawking for insects at the Upper Lobe, or when the juncos start trilling and the fox sparrows whistling, well before the equinox. Winter begins when the witch hazel blooms, when rafts of ruddy ducks appear on the Reservoir, when flocks of white-throated sparrows arrive and the saw-whet and long-eared owls appear in the Shakespeare Garden hemlocks and the blue spruce at Cedar Hill. Butterflies and dragonflies, crickets, and katydids define summer.

The Regulars notice what others have long learned to ignore: the sights and sounds, smells, textures, and tastes of the world around them. Forget the self and its hungry needs. Pay attention to tiny details. One wing bar or two? Six petals or eight? Listen to that squirrel whining. It probably means there's a hawk or an owl nearby. Notice the wind. In May if it's from the southwest, a wave of songbirds may be arriving. Pay attention to tree cavities. There may be a bunch of raccoon babies poking their heads out, or a family of young downy woodpeckers clamoring to be fed.

In their human relations, too, the Regulars' ways differ from the ways of the modern world. Neither job nor income nor family background confer a place in their hierarchy—nobody asks about these things. They don't matter. Among the Regulars, each person's skills and abilities count to secure the others' respect and approval. Among these are skills in observing, identifying, asking the right kinds of questions, skills, even, in encouraging others who might possess any of these skills to practice them. Is faithfulness a skill, an ability, a tendency, a trait? Whatever it is, it is admired, perhaps above all. The *regularness* of the Regulars is the feature that binds them together most powerfully. It allows them to count on each other to be there, observing, noting, keeping track. The Bird Register is their communications center.

ANYONE CAN WRITE in the Bird Register, and over the course of time many do. New birders, old-timers, out-of-town birdwatchers, ornithologists from the Natural History museum, tourists who want to express their delight with the park—all write occasional entries in the Bird Book.

During the spring and fall migration seasons, precisely when a greater diversity of birds shows up in the park, a diversity of contributors weighs in with entries. Names of legendary park birders appear then: Marty Sohmer, Michel Kleinbaum,

Peter Post—the Big Guns, as I think of them, who write only when they spot something out of the ordinary.

The Big Guns are actively searching for esoteric birds. While many bird enthusiasts will glance at a bunch of sparrows and dismiss them as ordinary house sparrows, one of the Big Guns will spend fifteen minutes inspecting this army of drab creatures, going over them one by one, and find in their midst a single, elusive Lincoln's sparrow. This bird resembles the song sparrow, a common park bird, but the side of its face and its eye stripe are a bit grayer, its breast stripes a bit finer, its bill a tiny bit thinner. Once Kleinbaum, a bird illustrator, pointed one out to me near Willow Rock. I always see better birds when a Big Gun is in the vicinity.

The Register would be a slim volume indeed if the Big Guns were its only correspondents. In fact it is a hefty volume by the end of the year. Most entries, as I saw that first day I looked in the Register, were written by a small number of men and women whose names appeared again and again—the Regulars.

TOM FIORE AND NORMA COLLIN are the Bird Book's principal correspondents. Both in their own way keep it a complete document of the natural and human history of Central Park.

Sun-darkened in winter as well as in summer, Tom Fiore reminds me of Thoreau's dictum: "A tanned skin is something more than respectable, and perhaps olive is a fitter color than white for a man,—a denizen of the woods."

Tom is the Bird Register's prize reporter. Until Tom began writing in the Register, no one had ever cataloged the park's daily and year-round birdlife (and other wildlife) in such detail. No one has been so faithful a daily chronicler in the years since.

Before 1991, the first year Tom appeared in the Register, most entries in the Bird Book simply named the species seen and the location of the sighting: "Fox Sparrow, Azalea Pond" or "Pintail, N.E. Reservoir." Tom's observations involved the mind as much as the eye:

> March 6: Four or more Fox sparrows near Azalea Pond, 2 of which seemed to be showing preliminaries-to-mating behavior: one calling and singing quietly, the other following, then both facing each other almost beak to beak and generally imitating what the other would do. This went on 10 minutes or so, then they continued feeding.

"Fantastic observations! Thank you!" Christina and Marianne Girards wrote in the Register that same day. They had been regular park birdwatchers since 1960, and welcomed Tom's more informative entries. They continue to come into the park almost daily, though Chris celebrated her ninetieth birthday a few years ago. Aunt and niece, the Girards are among the Regulars who bridge the pre-Tom and post-Tom gap.

The Regulars have come to depend on Tom's daily entries in the Register. Once, when this quietly intense, unfailingly courteous young man took off on a bird trip to the Philippines, the birdlife of Central Park suddenly seemed *so quiet!* It's not that the other Regulars aren't fine birdwatchers; it's that Tom seems to have a seventh sense that takes him to the right place at the right time.

Almost every other spring, for instance, Tom comes upon a migrating common snipe, a hard-to-find shorebird that everyone else seems to miss. But it's hardly luck that bags him the snipe. Thanks to his knowledge of its preferred habitat, in the middle of March Tom begins searching for the long-billed bird of bogs and wet meadows in the closest place to snipe paradise the park has to offer: the swampy, reedy area between Balcony Bridge and Bank Rock Bridge.

That snipe usually arrives during the third week of March, year after year. The eastern phoebe shows up at the Upper Lobe on March 13th, give or take a day or two. Most of the other migratory birds that use the park as a stopover seem to follow a precise timetable. They'll appear on their appointed date at the same place year after year. So that must be Tom's secret: he's learned these times and places, that's all.

Actually, that's not all. In addition to knowledge, Tom *does* seem to have luck. Once he was leaning back on a bench at the south side of the Harlem Meer, listening to a jazz band playing Duke Ellington, when he saw a bird of prey come zig-zagging from the east. Something about it caught his eye, and he grabbed his binoculars to give it a closer look. It was a Mississippi kite, a bird never before seen in Central Park (or even above it).

NORMA COLLIN MAY NOT enter the most sightings in the Register (though she writes in plenty) nor see the most elusive birds (though she is a fine birder), but she is the scribe, the town crier for the birding community. You can piece together the drama of the birdwatcher's daily life from Norma's distinctive vignettes. After reporting on fifteen species of warbler seen one morning in the middle of May, including the beautiful hooded warbler, she wrote:

> While enjoying good views of the Hooded, I looked up into the "lightning tree" and there was a raccoon having its midday nap. A tufted titmouse collecting stuff for its nest flew

> down to the raccoon and pulled a tuft of fur from its backside. The raccoon didn't stir, so the titmouse went back for another bill-full. This time the raccoon turned around and (if we can give it human traits) looked very surprised and slightly annoyed when the titmouse took a hunk of fur the third time. By that time around 8 birders had gathered around for the show and then went onward for a "Big" day of birding in the Ramble.

Though tufted titmice are known to gather fur or hair for their nests from live mammals, including dogs, squirrels, horses, cows, cats, and, amazingly, *Homo sapiens,* none of the texts includes the mammal that was titmouse-plundered in Central Park the following year: a red panda sleeping in an outdoor exhibit at the Central Park Zoo. The witness was Clare Flemming, a young mammalogist from the American Museum of Natural History. She subsequently reported her sighting at a meeting of the Linnaean Society, a scientific-minded club for ornithologists, birdwatchers, and nature-lovers that meets twice a month at the museum. Many Central Park birdwatchers are members and report their best park sightings at the end of the meeting when the president asks: "Are there any field notes?"

The entries in the Register resemble nature sightings in logs almost anywhere: Cape May, Point Reyes National Seashore, Muir Woods. Some of Tom's entries would not be out of place if they appeared in Henry David Thoreau's great *Journal:*

> March 11, 1854: Air full of birds,—bluebirds, song sparrows, chickadee [phoebe note] and blackbirds. Song sparrows toward the water, with at least two kinds of variations of their

strain hard to imitate. *Ozit, ozit, ozit, psa, te te te te te ter twe ter* is one. The other began *chip chip che we* etc. . . .

H. Thoreau

April 27, 1995—Weather: mild, s/sw wind, bringing light haze/sunny sky . . . Indigo Bunting, singing 6:20 a.m. Belvedere Castle (w. side), Red-eyed Vireo, singing, "Warbler Rock" (i.e. the trees on top of rocky outcrop, N.E. of Bow Bridge), Least Flycatcher singing: rapid Che-bek, che-bek call, Tupelo meadow area (early a.m.) . . .

T. Fiore

Yet there is the difference of a world between these observations. Thoreau described what he saw and heard in a natural surrounding unchanged for millennia. The world of Central Park is entirely man-made and its wilderness is enclosed by a city. Not even a city of well-tended gardens, with tree-shaded lanes and village greens: this is a city of skyscrapers on a bed of Manhattan schist.

"And imagine! This happened in the heart of New York City!" Nobody bothers to write these words in the Bird Register. Nevertheless, this context—the "other" world of nature within, civilization and its discontents without—informs each natural event in the park and deepens its excitement. The red-eyed vireo nest where both parents are feeding three nestlings is not hanging in a wooded glen; it is suspended over a path directly beside lamppost 7106, and as the parent birds go back and forth with large pale-green insects in their bills, legions pass directly underneath, some on foot, others on Rollerblades, bicycle, or tricycle, or in carriages or strollers pushed by parents, baby-sitters, or nannies in uniforms. And none of them have the slightest idea of the drama taking place inches above their heads.

On January 13, 1856, Thoreau found a red-eyed vireo nest at Walden Pond and wrote in his *Journal:*

> What a wonderful genius it is that leads the vireo to select the tough fibers of the inner bark, instead of the more brittle grasses, for its basket, the elastic pine needles and the twigs, curved as they dried to give it form, and, as I suppose, the silk of cocoons, etc. etc. to bind it together with!

His admiration for the vireo's genius would surely not have diminished at the sight of the nest at lamppost 7106, displaying bits of toilet paper, plastic wrap, and fishing line among the leaves and twigs and plant fibers.

ALL THE REGULARS keep an eye on the Register. Has it been returned to its proper place? Are the pages running out? Are people writing in the locations of sightings so others can find them? Are they signing their names? The names are important. News of the sighting of a boat-tailed grackle by one of the Big Guns would immediately activate the birders' grapevine, with one person calling another until everyone was reached. For this is a bird that has never been seen in Central Park.

The name of a beginning birder at the end of an extraordinary entry will evoke a cautious and mildly skeptical reaction. How many new birdwatchers seeing the common grackle, a large iridescent blackbird that abounds in Central Park, have looked it up in Peterson's *Field Guide* and identified it as a boat-tailed grackle, a *very* large iridescent blackbird. (Size can be misleading in a field-guide illustration.) The sighting will still be checked out; impossible birds do show up every so often, which is what gives birdwatching its special excitement.

Not long ago a new birdwatcher named Ilenne Goldstein wrote in the Register that she had seen a little blue heron at Turtle Pond, the small body of water at the base of Belvedere Castle.

"Please describe in *detail* the observation of the Little Blue Heron on 5/18 and please list any additional observers. This is *extremely rare* in Central Park!!!" wrote Tom Fiore under the notation. A beginner, he knew, might look in a field guide and mistake the rare little blue for the more common great blue heron, a bird often seen wading in the mud flats at the edges of that same pond.

Ilenne responded in the Register a few days later:

> He was an adult, completely blue body, reddish blue neck and head, bill dark at tip and silvery looking (to me at least) at the base. It was cloudy and muggy but the light was O.K. He had long reddish plumes on his neck that blew in the breeze and a "tassel" (also red-blue) on his head similar to a Black Crowned Night Heron's. He moved to a mucky area close to the south edge of the pond to feed alongside the mallards and sandpipers there. He was about the size of the mallards only much more slender. I watched for about 20 minutes. Unfortunately there were no other observers.

An impressive beginner, everyone agreed. When E. M. Forster proposed "Only connect!" as the basic creative principle, he was starting in the middle. Birdwatchers know something else must come first: Only observe. Take note of everything, *everything* (the dark tip of the silvery bill!) in its full, detailed richness.

In the Register the next day, Tom accepted Ilenne's identification of the little blue heron. Since no one else saw the bird in question, the sighting could never be finally confirmed for the record books. It was reported as a field note at the next Linnaean Society meeting, however.

EVERY SO OFTEN, in spite of everyone's tender care, the Bird Register disappears. It has vanished twice and had to be replaced in the six years I've been following it. Who in the world would want to swipe it? A tourist looking for a souvenir? Someone with a grudge against birdwatchers? A kid needing a new loose-leaf binder? Norma usually insists that it simply "got misplaced." Norma likes to think well of her fellow man.

The Register's disappearance inevitably brings on a feeling of dislocation, even desolation, among the Regulars. Like a family starting over after their house with all their old books and papers and *stuff* has burned to the ground, without the Register the Regulars have lost the history that tied them together. Now they seem no more than a collection of odd, disconnected people.

AS THE REGULARS study the natural world around them day after day, year after year, they confirm entomologist Vincent Dethier's astute observation: "Lack of knowledge is a vacuum that must be filled. Human beings are animals that must know." The Regulars share in equal measure a generous impulse to pass along what they have learned to others. One September morning when a weekly bird-study group, the Earlybirds, was walking by the Swampy Pin Oak, Dorothy Poole, a Regular and a superb birdwatcher, stopped and strained her ears. "Listen!" she said. "Do you hear that?" At first nobody did. There was just a big jumble of undifferentiated bird sound. "There it is again. That little 'whit, whit, whit.' " One by one we heard it, a tiny, repeated chip coming from the underbrush. "That's the Swainson's thrush," said

Dorothy. After a few moments we could hardly imagine we had *not* heard it before. We walked farther and again Dorothy stopped to focus our attention. " 'Vi-ew, vi-ew.' Hear how different that little chip is? That's the veery." The lesson was rounded out before long by yet another thrush chip, the quicker "pip-pip-pip" of the wood thrush. How had we come to ignore these sounds before, everyone asked. Had we been deaf?

NORMA COLLIN'S KNOWLEDGE of birds, plants and flowers, trees, fungi, butterflies, dragonflies, and other insects is extensive, though less encyclopedic than Nick Wagerik's. Nick, a superb birder and the park's preeminent butterfly and dragonfly expert, knows the various scientific names in the plant and animal kingdoms. Nick can draw you a perfect diagram of all the parts of a flower. The poetry of taxonomy is always at the tip of his tongue, as it was one day in the autumn when in the course of a walk in the Ramble he named and defined the various kinds of fruit in the plant kingdom: berry, pepo, hesperidium, drupe, pome, legume, follicle, capsule, silique, akene, caryopsis, samara, schizocarp, nut.

Norma may not know the textbook names of every flower, but she knows which ones are good to eat. In January Norma can show you where the field garlic is coming up; in April, the emerging young shoots of Japanese knotweed, the park's ubiquitous weed—they're quite tasty. In June Norma knows which mulberries are sweet and which taste like cardboard. (Their place in the sun matters.) Best of all, she knows where the secret wineberry bushes are and when they're due to ripen. Wineberries are more delicious than the finest raspberries, and the price in the park is right.

Norma knows where to find blackberries in August and persimmons in October. She once generously showed me the only fruit-bearing nectarine tree in Central Park, and gave me one of the "curious peaches" to sample. The tree's no longer there, by the way—done in by the drought of 1995. By September she'll have figured out which of the crabapples will produce the best fruit that year—it's not always the same ones.

And then there are mushrooms. A total of eighteen people shared in a mycological bonanza Norma discovered in Central Park one day in October—a decaying log covered with sulphur shelf mushrooms. Nick would have called them *Laetiporus sulphureus,* but Norma used one of their vernacular names—chicken mushrooms—because they taste somewhat like chicken when properly cooked. Everyone took some home and ate heartily. A sign of the special esteem in which Norma is held by her fellow Central Park nature lovers is this: it is common knowledge that some mushrooms can be deadly, but eighteen people put their lives on the line that day with perfect confidence. I'm still here to tell you that fresh *Laetiporus sulphureus* does taste like chicken, only a little better. (Her recipe: slice, sauté in a little oil, add a little water, cover, and continue cooking for about ten more minutes.)

NICK WAGERIK INTRODUCED me to the butterflies of Central Park on a sunny day in early June a few years ago.

Nick is a young man of medium height, solid, somewhat teddy-bearish in appearance. He lumbers when he walks, slightly hunched over. But the ursine appearance vanishes as he talks. The man is a bundle of nervous energy: His eyes dart about, taking in everything; words tumble out, a flood of information bursts forth. Nick is fiercely devoted to facts.

We were walking through a sun-dotted glade at the northwest corner of the Ramble that day when Nick pointed out a small butterfly fluttering by a rocky outcropping. "This one is an eastern comma," Nick told me, "*Polygonia comma.* It belongs to the family Nymphalidae. Commonly known as brush-footed butterflies. It's a woodland butterfly that loves glades with some sun breaking through. Like this one."

The butterfly settled on the rock, basking in the sun, and Nick took out his magnifying glass to show me a tiny silvery mark on its lower side (the ventral surface, as he called it). It did indeed look like a little comma, though I might never have noticed it if he hadn't pointed it out. Immediately afterwards I thought I saw another comma butterfly, but I was mistaken. "Look at its comma," Nick instructed. "Does it look the same?" As I peered at the insect through my binoculars, I saw that there was an almost imperceptible difference: just below the comma was a tiny dot. That dot made it a different species: a question mark butterfly, *Polygonia interrogationis.*

Nick possesses the indispensable characteristic of a great teacher: an infectious passion for his subject. As we were leaving the park after that first lesson in lepidoptera, he stopped and pointed out a pair of butterflies circling and swirling in the air just ahead. "That's the butterfly dance!" he said with emotion. "It's the most beautiful thing in nature. I could watch it forever."

ONE GREAT COMMON INTEREST—birds, or butterflies, or trees, or even natural history in general—might not be enough to transform a motley collection of individuals into a cohesive entity. The birdwatching community of Central Park is also a product of those events and occasions that periodically bring them together—gatherings of the tribe, you might call them. For just as weddings or funerals provide an opportunity for members of a far-flung family to meet and reaffirm membership in the group, so these occasions serve to define the birdwatching community and to cement its rather amorphous bonds.

There are organized events: the Christmas Count, the fall migration Hawkwatch at Belvedere Castle, the occasional celebrations to mark special birthdays—usually avian ones. These are the national holidays, so to speak, of the birdwatching community, marked on all calendars well in advance.

Then there are the serendipitous gatherings of the clan whenever a rare or special bird is sighted in the park. Engraved invitations will not do for these get-togethers—time is of the essence. Hence the birders' grapevine. Who will forget the January morning when a four-star bird for Central Park, a great horned owl, was discovered on the outskirts of Muggers Woods. (In spite of its name, the little thicket just behind the Humming Tombstone attracts no more muggers than any other part of the Ramble.) Tom Fiore rushed for the phone at the Delacorte Theater, while Charles Kennedy, one of the most faithful of the Regulars, biked to the outdoor booth at the Boathouse. Each made three or four calls whose recipients made still others. On their way back to the owl, Tom and Charles passed the word to Regulars already in the park. By the end of the day a total of twenty-two people had seen the great horned owl, most of them within an hour or two of its discovery.

ACT I

The Odd Couple

The birdwatcher's life is
an endless succession of surprises.

W. H. HUDSON

SCENE ONE

Enter Pale Male

❧

The name of the game in birdwatching is telling one species of bird from another. This one's a chickadee; that one's a nuthatch. But only by marking birds in some way, usually by attaching bands to their legs, can individual birds within a species be distinguished from each other. Generally one robin looks like any other, or one downy woodpecker, or one red-tailed hawk.

Every so often it happens that a particular bird displays some feature that makes it recognizable as an individual. Sometimes there's an injury—a duck with a broken wing; sometimes a genetic anomaly—an albino sparrow; sometimes it's simply a natural oddity of size or shape or color. People may follow the course of such birds' lives in a way that would be impossible if they looked just like their species-mates.

Such a one was the red-tailed hawk that arrived in Central Park during my first winter as a Regular. He had a feature so distinctive he could always be identified—not just as a red-tailed hawk but as himself, a particular, individual bird. Whereas this species appears in field guides with a white breast, a broad band of streaks across the belly, and a darkish head, this particular red-tail was exceptionally light all over. His head was almost white. He had no belly-band to speak of—the breast and belly were white. He wasn't an albino, his eyes were too brown; just a very pale red-tail.

Tom Fiore saw him first, and reported his sighting in the Bird Register on November 10th:

> There is a very light-colored, immature red-tailed hawk that has been seen eating a rat and also swooping a foot above shoveler ducks on the lake.

How did Tom know the light-colored hawk was immature? Was it smaller?

It's a common misconception that all young birds are smaller than adults of their species. Though it is true for ducks, geese, and birds that leave the nest shortly after hatching, most songbirds and birds of prey, indeed, the majority of the avian order, cannot leave the nest until their flight feathers grow in. These birds are full adult size by the time they take their first flight.

It is the plumage of most immature birds, not their size, that clearly distinguishes them from the adults. Young birds generally lack whatever bright colors adults of their species might display, making them less conspicuous during the vulnerable nestling and fledgling periods. Young robins have speckles on their breast rather than the characteristic solid brick red coloring of adult robins; juvenile male cardinals are buffy brown, not bright red, as their fathers are. Immature red-tailed hawks contradict their name: their tails are brown, and won't turn red until they are two years old—breeding age. Moreover, the tails of immature red-tails are marked with distinct black bands. One look at the pale-colored red-tail's tail revealed to Tom that this one was a kid.

For a while Tom was not sure whether the young hawk he'd sighted on November 10th was a male or a female. Among most birds of prey, males and females do not significantly differ in plumage or markings. Still, there *is* a way to tell the sexes apart: Female raptors are almost always bigger than males. Consequently, when a second and noticeably larger hawk arrived in the park a few months later, the sex of the pale-colored bird became perfectly clear. The Regulars began to call him Pale Male.

He was still a browntail that spring, too young for love. Nevertheless and notwithstanding, when the female (this one with a bright red tail) showed up at the beginning of March, Pale Male courted and won her.

❧

THE FIRST JUNCO trill was noted by Norma Collin on March 9th that year, right on schedule. One week and one day later Pale Male and his new mate made their decision. They'd been checking the place out for a few months now and things looked good. Plenty of food around—corn-fed pigeons and garbage-fed rats. A lake to bathe in. Protection available from wind and storms. Time to get the show on the road. On March 17th the hawk pair began to build a nest in Central Park. It was a historic event, for in the 119 years of the park's existence, no hawk had ever nested there before.

They made an odd couple, a mature female who had hooked up with a young and inexperienced male. His lack of savoir-faire was evident: on March 23rd, as the female was perched on a stanchion in front of the Delacorte Theater below Belvedere Castle, the light-colored red-tail was observed as he landed on top of her and tried to consummate their union. But he was doing it at the wrong end.

Perhaps lack of experience also explains the absurdly conspicuous spot the birds chose for that historic first home in Central Park: a tree just behind the baseball backstop at the southwest corner of the Great Lawn.

A grassy expanse filled, on nice afternoons and weekends, with kids playing Frisbee, muscular men in shorts and knee socks playing soccer, and various groups enjoying pickup games of baseball, the Great Lawn is hardly an auspicious nest site for a bird long thought to be highly sensitive to the human presence. "It is one of the shyest of our hawks," wrote Arthur Cleveland Bent in *Life Histories of North American Birds of Prey.* "If [red-tails] suspect that the nest is watched they will not come near it." That was in 1937. The species had obviously adapted to new circumstances over the years, for these Central Park red-tails were oblivious to the hundreds of humans in the vicinity. Most of those humans were equally unaware of them.

The park's birdwatchers, however, took notice. They watched the hawks bringing sticks to the nest. They timed the increasingly successful though invariably brief love acts. They saw them hunting for pigeons and making occasional dives at local squirrels. Accustomed to receiving handouts in perfect safety, the Central Park squirrel corps was totally unprepared for becoming handouts themselves for resident birds of prey. And as the first migratory birds of spring, the woodcock and the eastern phoebe, began to trickle into the park, the Regulars saw that these much-awaited visitors had something new to contend with. As Tom Fiore noted in the Register on March 25th:

> While bluejays in nearby trees were screaming . . . the pale [immature] Red-tail came flying up from those trees and landed on a mat of broken phragmites in Belvedere Pond. . . . I sat down at the edge of the rocks below the Castle and a Woodcock flew from behind me right over the young hawk,

> toward the north shore of the pond . . . the young hawk flew right up and chased after the woodcock, pursuing it through some trees on the south shore. . . . an Eastern Phoebe was catching something too small to see, moving around the area below the Castle. Then the young Red-tail went after *it,* but the Phoebe merely landed on a little outcrop of rock below the Castle and the hawk was forced to veer off.

On April 2nd Norma Collin's Register entry made it clear that Pale Male had acquired some new skills:

> Red-tailed hawks *mating* on Delacorte tower. Mature female (with red tail) stayed near nest. Male circled and returned and they mated again!! So keep watching!

But by April 4th it was obvious that the Great Lawn nest was a bust. The flimsy structure was falling apart. That was the day Sylvia and Mo Cohen, birdwatchers who go back to the old Lambert Pohner days, found a broken egg at the base of the backstop. They brought it to the next meeting of the Linnaean Society, where one of the club's officers examined the fragment and said: "This proves that anything's possible in Central Park."

Less than a week later, a Register entry revealed that the indomitable pair had not given up.

> There's another *nest* being built by Red-tailed hawks in the park. I'm not saying where it is for now. Perhaps you've already discovered it?
>
> Tom Fiore

Tom was clearly worried that birdwatcher attention may have contributed to the failure at the Great Lawn. Perhaps secrecy would help, at least in the early stages of nest building, he decided. But two days later someone else described the new nest site in the Register:

> Red-tail in nest! East drive, slightly north of marker 6902—west side of Drive—(elm).
>
> Murray and Dave

On April 14th Tom relented and wrote in a detailed description of the new nest site:

> Red-tailed Hawk on *nest*—high in (Elm?) tree—large, dark bark with deep furrows. It's between 2 large London Plane trees just 15 feet or so west of the East Drive (opposite E. 70th Street). There are daffodils planted nearby and a new water fountain on the path just west of the nest tree.

The next day Tom drew a map of the nest location in the Register, though he still worried about the consequences of too much human attention. "Please don't make a lot of noise if the hawks are seen," he wrote in large letters beside the map. It seemed an unnecessary injunction since the birds' new location was directly adjacent to SummerStage, a park facility where noisy crowds often gather and ear-shattering rock concerts are held on summer weekends.

A group of Regulars began gathering daily to watch the birds at the nest. By the end of the second week of April the large structure of sticks and twigs near the crown of the tree seemed complete. This time, the nest looked more solid. Everyone made efforts to be quiet in the vicinity, and those who forgot themselves were "shushed" into submission. But had the hawks built a home as strong as Fort Knox and had they enjoyed a human audience as silent as the grave, their efforts would still have been doomed. They were done in by members of their own class: crows.

Crows are natural enemies of hawks, and little mobs of them are often seen in furious pursuit of birds of prey. Yet these crows seemed exceptionally vicious. From the first day the hawk pair tried to settle in at their new nest site, the entire crow nation seemed to have declared total war. There proved to be a reason for the ferocity of the crows' attacks: they had a lien on that territory, for the previous year a pair of crows had nested in the same elm. Now, as the birdwatchers watched the screaming squadrons of crows persecuting the hawks, they lavished pity on the nesting pair, quite forgetting that hawks have a savage side of their own.

The hawks persevered, and by the end of April their behavior made it clear that there were eggs in the nest. One hawk would arrive, the sitting one would take off, whereupon the arriving bird would settle down to keep the eggs warm—nest exchanges, these are called.

The crows persevered as well. At each nest exchange, the crows hounded the departing hawk, their screams ever louder and more insistent. The end came on May 2nd, when the female hawk was so flummoxed by her crow persecutors that she crashed into a wall at the top of a high-rise on Fifth Avenue and 73rd Street.

Witnesses in an adjoining building called the New York City Audubon Society, which in turn called a local wildlife rehabilitator, Vivienne Sokol. But when she arrived it was clear that the injured hawk was beyond first aid. Something was seriously wrong with the bird's left wing, which hung at a weird angle to the side. Miss Sokol packed the bird into a large cardboard carton and took her away.

A few days later a particularly vociferous gang of crows, perhaps emboldened by their successful rout of the female hawk, pursued Pale Male until he too crashed into a building, this one at 62nd Street near Madison Avenue.

A woman I ran into by chance a few years later told me the whole story. She then worked for an ear-nose-throat doctor in that building, a Dr. Blaugrund. At around noon on May 4th she heard crows screaming bloody murder just outside the waiting room. She ran to the window and instead of crows she saw a large brownish bird lying on the rooftop of a town house below. She too called the Audubon Society, and then all work stopped at the doctor's office as everyone—doctor, nurses, and patients—stood by the window, watching the unconscious hawk. Things did not return to normal until the wildlife rehabilitator arrived, the same one, it turned out, who had taken charge of the injured female.

Pale Male had merely suffered a concussion. Miss Sokol applied first aid and gave him a few days to recover at her home. The female hawk was beyond local help. The rehabilitator packed the bird back in a large carton and drove her to the Raptor Trust, a well-known hawk hospital in Millington, New Jersey. If anyone could help the bird it would be Len Soucy, the founder and director of the facility.

The Regulars were stunned at the sudden conclusion of the red-tail affair. It had been a thrilling story. But everyone knew that other stories were happening elsewhere in the park. They just had to keep their eyes and ears open. And at least Pale Male was back in the park. "Hallelujah," said Charles Kennedy, a man given to exultation and to looking on the bright side of dismal events.

SCENE TWO

Heartbreak in the Ramble

Each spring the arrival of the Baltimore oriole in Central Park is an occasion for rejoicing among the park's Regulars. Not only is it one of the handsomest of songbirds, not only does it have a particularly melodious song, but unlike the warblers, the tanagers, and the hundreds of other migratory "tourists" for whom New York is "a nice place to visit (twice a year) but wouldn't want to live there," the Baltimore oriole considers the park a fine place to settle down and raise a family. True, unlike the downy woodpeckers, blue jays, cardinals, and other permanent residents, the Baltimore oriole doesn't stay the year round in Central Park. Nevertheless, it is as much a New Yorker as all those folks who live in the city but winter in Florida. A Regular.

Baltimore orioles spend their winters south of the border, in Central or South America, fattening up and preening their colorful plumage—bright orange and black for him, olive and yellow or drab orange for her. In April they head north for their breeding grounds throughout eastern and central North America. Like so many migrants of all classes, they leave home for economic reasons: too much competition for food, too few housing opportunities. Male Baltimore orioles arrive a few days earlier than the females in order to establish their breeding territories. Getting there first means getting a good nesting spot.

Traveling in the company of huge numbers of other migrants, orioles fly by night at a height of between 1,000 and 2,000 feet. During the daylight hours they stop to rest and feed at various green spots along the way. Why do they fly by night? The most widely accepted explanation for the phenomenon of avian night migration used to be predator avoidance: by flying at night, small birds obviously reduce their chances of becoming hawk food. But ducks, geese, and egrets, birds too large for predator attack, also migrate at night—making this a less plausible explanation. Another theory proposed that the stars are an important aid for navigation. But recent findings show that there are many cues available in the daytime, visual landmarks such as rivers and mountains. The earth's magnetic force and polarized sunlight help to guide daytime migrants as well.

In his book *How Birds Migrate,* ornithologist Paul Kerlinger summarizes today's preferred explanations for night migration. Temperatures are always cooler at night—that, according to Kerlinger, is the decisive factor. Birds flying in higher daytime temperatures lose more water than those flying by night, water required for thermoregulation and for metabolism. During long and necessarily uninterrupted flights over oceans or deserts when they cannot replace lost water, migrants run the risk of dying of dehydration. By flying at night they can go greater distances on smaller water reserves. In addition, the absence of updrafts associated with daytime thermal currents allows night-flying birds to maintain straight, level courses. Thus they can conserve precious energy on long flights.

Even with these strategies, developed over the millennia, great numbers of migrants perish on the way. But far greater numbers, millions of birds, manage to reach their destination.

For five or six pairs of Baltimore orioles, that destination is Central Park.

AT THE VERY TIME Pale Male and his mate were sitting on eggs in the big elm near SummerStage, a small bunch of flashy male Baltimore orioles arrived in the park. They showed up at the southern end on April 26th—right on schedule. There they dispersed and staked out their territories, singing and chasing each other around until all the boundaries were established. The hero of this story was among them. He immediately headed for the Cherry Hill area, an insect-rich location near the rowboat lake.

Orioles are a philopatric species—they keep coming back to the same nesting place year after year. Odds are that the Cherry Hill male had nested in the vicinity before. He soon staked out his claim to the area around a large, hybrid poplar tree standing at the water's edge near lamppost 7305. The spot is located between two park landmarks: Bow Bridge and the Bethesda Fountain.

The females arrived a week or two later, and for reasons no more or less mysterious than the choice of mate in our own species, the Master of the Hybrid Poplar fancied a particular female. Courtship began forthwith. First, as is the custom among orioles, he chased her as if she were another intruding male. Then he proceeded to endear himself by performing the bowing ritual. As described in Donald and Lillian Stokes's *Guide to Bird Behavior:* "He lands in front of the female, bows low, jerks up, and then bows low again." Writing in his great work *Birds of Massachusetts,* ornithologist Edward Howe Forbush described the courtship in more detail:

> In displaying his charms before the object of his affections the male sits on a limb near her, and raising to full height bows

> low with spread tail and partly-raised wings, thus displaying to her admiring eyes first his orange breast, then his black front and finally in bright sunlight the full glory of his black, white and orange upper plumage, uttering, the while, his most supplicating and seductive notes.

Who could resist such courtly wooing? The two became a pair.

All during the staking-out-of-territory stage and the courtship phase, the male oriole sang loudly and clearly from the treetops. The song is easier to recognize than to describe. Thoreau transliterated it with the words "Eat it, Potter, eat it!" Another bird expert, T. Gilbert Pearson, distinctly heard an oriole in Pittsburgh singing "Ta-ra-ra-boom-de-ay!" Pearson found the oriole's song "singularly cheerful," and in 1996 the Baltimore oriole finally had something to be singularly cheerful about: on the basis of irrefutable DNA evidence, the American Ornithologists' Union (which is in charge of such things) declared that it was a separate species from the Bullock's oriole, with whom it had been lumped for twenty-three inglorious years under the catchall name northern oriole. (Luckily, the baseball team never changed *its* name to the Northern Orioles.)

Next commences the most remarkable stage of the Baltimore oriole's breeding cycle: nest building. The female constructs the nest alone, as is the case among other blackbirds (the oriole is a member of the Icteridae, or blackbird family), hummingbirds, tanagers, and finches. In other species, males do their bit by bringing construction materials for the female to use or building certain parts of the nest, though generally, among birds as well as among many mammal species, females are in charge of preparing the place where the young are to be raised.

Known as the most skillful artisan of any North American

bird, the female oriole weaves a long pouchlike nest out of dry plant fibers, hemp, milkweed silk, and other available materials. Feverishly she works for five to eight days, shuttling back and forth, pushing the fine threads through the nest body and drawing them through, thrusting and pulling, working from above, then from below, balancing herself with spread wings, until the job is done. The nest is firmly attached to the very tip of a delicate, flexible branch. This Darwinian strategy protects the nestlings from predators too heavy for the branch to support, such as raccoons or squirrels. Plenty of both in Central Park.

The lakeside nest was completed on May 21st, the eggs deposited shortly thereafter, whereupon the fourteen-day incubation period began. Again the female took on the job alone, sitting deep in the pendant nest. Only the very top of her head was visible from below. The male kept guard, serenading her with song (more likely he was warning off potential rivals). He brought her occasional tidbits to keep her going.

On June 6th the eggs hatched. Though the nestlings were completely hidden from sight and would remain so for several days, the birdwatchers knew they had hatched by the parents' behavior: They had begun shuttling back and forth to the nest with food for the babies.

On the first two days the parents carefully predigested various insects and larvae before feeding them to their featherless young. On the third day the parents could be seen removing a dragonfly's wings before entering the pouch.

On June 11th, somewhere between 9:00 a.m., when Dorothy Poole passed by on her way to work and all was well (the nestlings must have feathers now, she thought), and 10:00 a.m. that same morning, when Norma Collin came to check out the oriole family, the entire tree vanished. When Norma arrived all that remained was a huge stump.

Norma Collin is not a crack-of-dawn birder like Tom Fiore. She starts her day slowly and likes to have a leisurely bowl of oatmeal or hominy grits before heading for the park. When she arrived at Cherry Hill it was all over; a demolition crew had just finished stuffing the last branches of the poplar into a tree shredder. Norma's entry in the Bird Register that day did not hide her feelings:

> BAD NEWS! The Poplar (or cottonwood) tree in which the orioles were nesting was *cut down* at 9:00 a.m. this morning. When I got there at 10:00 a.m. the Parks Department had already mulched the greenery and were loading the truck to be hauled away. No sign of the nest, but the adult male and female orioles were flying around with bugs in their bills looking for the nest and landing in the bare branches of the cherry tree to the west, and the ginkgo tree to the east of the site where the poplar tree had been.

"It was heartbreaking," she said to a group in the Boathouse later that day. "The parent birds were flying around in confusion looking and looking for the nest with the babies. If only I had come earlier!"

Word spread rapidly, and reaction was swift: outrage.

"How would I feel if I came home and found my house gone, my kids vanished, in fact my whole block evaporated?" That was the immediate response of Rebekah Creshkoff, a young woman then becoming one of the park's most accomplished birdwatchers. She did more than empathize. She took action. From her downtown office at Chase Bank, she telephoned Neil Calvanese, the park's director of horticulture. He makes the final decisions about tree removals.

When Calvanese answered her call, Rebekah did not mince words. Didn't they know they had shredded a nestful of oriole babies? Calvanese, a nature lover, was genuinely appalled. He replied that, indeed, they hadn't known. The tree was half-dead, ill with a canker disease, and had to go. Well, chided Rebekah, unmollified, they *should* have carefully checked the tree for nests before cutting it down.

After Rebekah had applied the moral screws to the New York City Parks Department, she called the U.S. Fish and Wildlife Service and discovered another compelling reason for checking a tree before feeding it into a shredder: the Migratory Bird Treaty Act of 1918. This law declares that "it shall be unlawful at any time, by any means or in any manner to pursue, hunt, take, capture, kill . . . any migratory bird, any part, nest, or egg of such bird."

As a result of a general outcry of park bird lovers—telephone calls, letters to the Parks commissioner, and, perhaps not least, a reminder about the Migratory Bird Treaty Act—the Parks Department instituted a new policy. Before cutting down any tree, Mr. Calvanese declared, henceforth they would call in a park ranger who would check the tree for nests and birdlife. "This was an unfortunate accident. We're committed to making sure it doesn't happen again," Mr. Calvanese said.

The Regulars weren't taking any chances. They too were committed to making sure it didn't happen again. And so they formed a Nest Patrol under the leadership of Rebekah Creshkoff. Later that year it informed Mr. Calvanese of a warbling vireo nest in a rather sickly Norway maple just across the path from where the hybrid poplar had once stood. The tree remained in place for two more years. Then it, too, had to go, but not until the late fall, when nesting season was safely over for even the latest nesters.

SCENE THREE

Feeding the Birds

When it's horribly cold in Central Park, when ice and snow cover the ground and arctic winds rage through the bare ruined choirs where late the sweet birds sang before most of them took off for warmer climes, then tourists or dog walkers passing by the feeding station stop and inquire: How do the remaining birds survive? Why don't they all freeze in the winter?

Birds, in fact, have a remarkable resistance to cold, far greater than mammals do, according to John Terres' *Audubon Society Encyclopedia of North American Birds.* It makes sense, when you stop to think of what many mammals of our own species wear when the weather turns wintry: down garments. Birds, of course, are protected by their own natural down. Feathers, in fact, are one of nature's finest insulating agents.

There is very little exposed skin on the body of a bird—no projecting ears or tail where body heat might be dissipated. Only the feet and, to some degree, the horny bill are vulnerable to heat loss, which explains why you sometimes see a bird standing on one foot with the other tucked into the breast feathers, or with its bill buried in a wing to conserve heat.

As winter approaches, the factor that impels birds to fly

south is not the cold. It is the food supply. The birds that migrate depend on insects for their primary food source. When the weather turns too cold for insect survival, these birds must move to where the food is. The birds that stay—the chickadees, titmice, jays, woodpeckers, crows, mockingbirds, finches, and sparrows—are generally those that can survive on food available on the ground or clinging to leafless branches during the winter—seeds, dried berries, and the like. Birds such as woodpeckers and brown creepers have beaks adapted to ferreting out insect larvae buried deep in the bark of trees.

When snow and ice cover the ground and shrubbery, the food supply is hard to get at. Yet those are the very times that birds must eat greater quantities than usual to sustain their normal body temperature, which fluctuates between 100 and 112 degrees. It is surely a common-sense understanding of these realities that underlies the widespread practice of feeding birds in winter.

NORMA COLLIN, Charles Kennedy, and Murray Liebman, an English teacher at La Guardia Community College, form the nucleus of Central Park's informal bird-feeding squad. During the feeding season—from October or November (depending on the weather) to the middle of April—they meet every Wednesday at the Ramble Shed, a storage building near the East Drive and 76th Street. That's where the seed and the bird-feeding paraphernalia are stored. From there the three Regulars—and anyone else they can dragoon into helping them—lug the heavy bags of seeds and an extra-long collapsible pole to the Azalea Pond.

Located deep in the heart of the Ramble, the Azalea Pond bird-feeding station gives the impression of being an official part of the park's operations. Far from it. The Regulars run it on their own, accepting voluntary contributions to cover costs and answering to no one but themselves and to the songbirds, pigeons, hawks, squirrels, rats, raccoons, and occasional dogs: the customers.

Seven or eight outsized contraptions hanging high in the branches of trees—the feeders—attract the eye at once. Cleverly fashioned out of plastic gallon jugs, they are attached to branches by means of coat hangers twisted into hooks. Birds may enter through a small hole cut into the sides. A ⅜-inch dowel inserted just below the hole serves as a perch, while a Frisbee placed between the hook and the jug keeps squirrels out.

The feeders look like they might have been designed by Rube Goldberg. In fact, they are the creations of George Muller, a longtime Regular. "Many moons ago," as he puts it, he was an engine-room wiper in the Merchant Marines. That's where he observed a tin collar attached to the mooring line of docking ships. This device, which served effectively to keep rats from getting aboard, inspired the Frisbee on the bird feeders.

Every passerby asks the same question upon first spotting the feeders hanging fifteen feet or more above the ground: How do they get those big things up there? Another ingenious invention does the trick—a telescoping pole made out of hollow electrical conduit pipes of varying diameters. When all the pieces are put together it is fourteen feet long. For the very highest hanging feeder, yet another extension is added to this pole, a six-foot-long stick kept hidden behind one of the azalea bushes at the pond's edge. In addition to the hanging feeders, there are several rectangular wire gizmos filled with raw-beef suet—a staple of most birds in the winter. The suet feeders are attached to high branches by means of the same telescoping pole. Finally there are the thistle socks, hanging from the same branches and filled with the fine seeds preferred by finches and siskins.

Joe Richner, a retired freight inspector and weighmaster for the erstwhile New York Central Railroad, was a dominant member of the feeding squad until his death at the beginning of 1996. He was not quite eighty that year, and had been feeding birds in the park for over thirty years. Suet was Joe's specialty. All winter and spring, and even in the summer when the regular feeders were taken down, cleaned, and stored (and when plenty of natural food for birds is available throughout the park), Joe would bring in big chunks of the flaky white stuff to put up for the woodpeckers. It kept the beautiful creatures coming, to be admired the year round. Joe also took a lively interest in other aspects of the feeding operation.

"In the old days, people just put up sunflower seeds and not millet," he recalled. "But millet attracts sparrows, house finches, pine siskins, goldfinches, purple finches, redpolls, and also grackles. I introduced millet. And when the goldfinches came, I always put up finch seed, which cost four dollars a pound. We had two goldfinches that wintered here for two years. And for three or four years we had a wintering hermit thrush. I gave it apples and raisins I soaked overnight for him. He also ate suet."

Joe was usually there when the feeding squad arrived on Wednesdays. He would start talking immediately. "When I meet someone, I always say a few words to them. A few thousand words, that is," he liked to say. Lack of self-awareness was not one of his failings.

The feeding squad often meets other people who cater to birds—independent operators, one might call them. The winter I first began dropping by the Azalea Pond regularly Bill DeGraphenreid was solidly ensconced there.

Bill liked to throw peanuts to titmice, blue jays, and cardinals. "Here. Get this," he'd say firmly, and toss a nut directly in front of a bird. His aim was remarkable, and the bird often did get it. Until he moved his feeding operations to another location, many of the birds that frequent the Azalea Pond seemed to recognize Bill. But so did the squirrels. They'd gather from far and wide when they saw him settling down at the rustic bench, and the bushy-tailed cadgers often got the peanuts he was tossing to the birds before the birds did.

Bill was the park's preeminent chef for birds and usually arrived with a home-cooked treat—gruet—named for its two main ingredients: hominy grits and suet.

"Come and get it," Bill crooned as he pressed the greasy glop into the bark of the deeply corded cork trees that dominate the Azalea Pond feeding area. Though their botanical name, *Phellodendron amurense,* derives from the Greek word for cork, and though their bark does have a soft, corklike texture, these are not the trees wine-bottle corks are made of. That honor belongs to the cork oak, *Quercus suber,* a native European tree not found in the park.

Within minutes the red-bellied woodpeckers would arrive, announcing themselves noisily with their characteristic "Chork!" A couple of smaller downy woodpeckers might check in shortly thereafter, followed by a procession of titmice and nuthatches, birds that usually frequent the hanging feeders. They line up for gruet when it's available. The suet concoction is also a clear favorite of the tiny brown creeper, a well-camouflaged relative of the titmouse clan, and the only representative of the Certhiidae family in all of North America.

I didn't think Bill would ever share his secret recipe with me, and, in fact, when he finally relented, he wouldn't let me write it down. "Just listen!" he chided. "Stop writing." But he had me repeat the recipe step by step when he finished, and I wrote it all down the moment I was out of sight.

I kept talking about making gruet and I kept putting it off until a snowstorm struck toward the end of January. Bill, who has trouble with bone spurs and uses a cane, couldn't come in for almost a week. Everyone was worried about the birds, and it was clear that the time had come to put my money where my mouth was.

My first venture as a bird cook was harrowing—rendering beef suet is an arduous task, and I had never made hominy grits before. Bill always emphasized that the texture had to be just right, otherwise it wouldn't stick to the tree properly. "Smells great around here," my husband said when he came home that evening. I was forced to admit that the delicious dish he smelled was for the birds—our dinner was Chinese take-out.

I put up my gruet the next day and the first customers were two red-bellied woodpeckers, surely the most spectacular of the park's common birds. (They became hoi polloi birds only in recent years, being a southern species that has expanded its range. A few decades ago they were unheard of in Central Park.) These large black-and-white-striped birds with bright red caps and no red bellies to speak of—in breeding season there may be a faint blush visible—were quickly followed by three downy woodpeckers, two nuthatches, four titmice, a pair of chickadees, a brown creeper, and a white-throated sparrow, even though this species usually feeds on the ground. My creation enticed this bird to eat up on a branch of a tree—a mad success, as the French say.

ONE BITTERLY COLD January day a few years ago the bird-feeding squad set forth on its mission of mercy with a particular sense of urgency. The city was enduring one of its worst cold waves ever and there were already reports of bird victims: a frozen mallard had been found in the ice between the Loch and the Pool, the two water bodies of the northwest end of the park, and a dead Canada goose on the north side of the Reservoir.

The temperature was below zero that day as Norma, Charles, and Murray, three latter-day St. Francises, cautiously made their way along the icy paths to the Azalea Pond. But even the bitter cold couldn't numb the shock that awaited them at the frozen pond.

The huge feeding containers were a shambles, with gaping holes where the small openings had once been. They were hanging crazily askew, swaying in the breeze, completely empty.

The agents of destruction were not hard to identify: squirrels. The desperate rodents, their buried supplies made inaccessible by the snow and ice cover, had finally penetrated the defense system. They had twisted the Frisbee baffles, gnawed through the bird-sized entrance holes in the jugs, eaten all the seeds, and, in the process, damaged the containers beyond repair.

The squad had lugged a twenty-pound bag of birdseed to the feeding station, as well as the collapsible pole. Now, as they numbly wondered whether to simply dump the seeds on the snow-covered ground and let the birds and squirrels fight it out among themselves, another angel of mercy appeared on the spot.

It was George Muller. He too was amazed at the squirrels' depredations. But George had come prepared. After checking the Azalea Pond the previous Friday and seeing the squirrels at work, he emptied out a gallon jug of Breck shampoo he had at home, and appropriated another large plastic container he normally used for watering his plants. A third jug he commandeered from his building's basement. He cut out bird entrance holes at the sides, inserted dowels just below, and attached snap-hooks to the top of three new feeders.

And so it came to pass that on the coldest day of the year George, Murray, Norma, and Charles filled three new bird feeders to the brim with sun-flower seeds and hung them on the branches of an overhanging cork tree. They filled Joe's suet feeders with two large globs of suet. Then, in the spirit of true forgiveness, they scattered quite a quantity of additional seeds for the hungry squirrels.

As soon as the feeders were up, the customers began to arrive. Two downy woodpeckers were first, announcing their arrival with loud, descending whinnies—that must have been the dinner bell. Soon titmice and chickadees were flying back and forth from the feeders to nearby branches: Dee-dee-dee, they called as if in gratitude. On the ground two kinds of sparrows, white-throated and song, together with a large contingent of house sparrows—which are actually not sparrows but weaverbirds from Eurasia and North Africa—all chirped as they fed on scattered birdseed alongside a considerable mob of rock doves (pigeons, that is). And in the middle of January, deep in the wilds of Central Park's Ramble, the bare ruined choirs resounded with sweet bird song.

Postscript

Central Park, the same winter

Dear Holden Caulfield,

All through *The Catcher in the Rye* you kept asking a question, a really good question, and nobody ever gave you an answer. You remember—the one about the ducks in Central Park. You were worried about what happens to those ducks when all the lakes freeze over. You wondered whether some guy came in a truck and took them away to a zoo or something. And then in Chapter 12 all Horwitz the taxi driver would say was, "How the hell should I know?"

Well, all these years I've kind of wondered about your question myself. But not very hard. Because things have changed since the 1940s or '50s when you were meeting old Sally Hayes under the clock at the Biltmore. Would you believe it, Holden, that clock isn't there anymore. They've taken down the whole hotel, for goodness sake!

But that's not what I mean when I say that things have changed. I mean that it hasn't been so cold these winters.

I remember Central Park in the '40s and '50s. In those days it got really cold in January or February. Kids used to actually ice-skate on the old rowboat lake.

In those days I didn't worry about the ducks like you did, Holden, I honestly never gave them a single thought. But by the time I grew up and began to really care about ducks and stuff, the winters stopped being so cold. I don't know why, exactly, maybe the greenhouse effect or whatever. But it's the truth. The lakes in Central Park hardly ever froze over during the last few decades, not solidly so you could skate on them, and not all over so anybody had to worry about the ducks.

But recently I've been thinking about you a lot, Holden. Because this has been one unbelievably cold winter. I mean it's been *really* cold. All the lakes and even the Reservoir in Central Park have frozen solid. People are skating on the rowboat lake, for goodness sake. So your question began to really bother me.

And guess what, Holden, I actually found out what happens to the ducks in Central Park when everything freezes over. And I can tell you that nobody comes with a truck and takes them away to the zoo or anything.

My friend Bill DeGraphenreid found them. He's this nice black guy with a big shock of white hair who feeds the ducks all year. And imagine this: he actually knows those ducks. I'm not kidding. There's this one female mallard he calls Missy, and there's all Missy's children—she had eight ducklings last spring—and there's Missy's sister who was slightly crippled from getting tangled in fishing line. When he calls "Missy, Missy!" one of the Missys always comes.

Anyhow, when all the lakes froze this year, Bill began to worry about the ducks. So he looked all over the park for them. Finally, he found them. All of them, including Missy and Missy's sister. They were all in a secret place, just about the only place in all of Central Park that hadn't frozen over. Because there's an actual *natural* spring that runs into it, while all the other streams in the park turn on and off with a faucet, for goodness sake.

So Bill's been going there just about every day and feeding Missy I and Missy II and all the other ducks heaps of food, even though the roads in the park have been horribly icy and he's had this painful foot condition, bone spurs, that makes it hard to walk.

So Holden, I'm going to tell you how to find the secret place where the ducks in Central Park go when all the lakes are frozen over. Do you know where Balcony Bridge is? It's this structure that is actually a part of the West Drive, somewhere around West 77th Street. If you stand on its east side, you get a fantastic view of the rowboat lake and the Central Park South and Fifth Avenue skylines. From its west side, you're facing the Natural History museum.

Well, all the ducks are right there and nobody hardly notices them. Hundreds of ducks right under old Balcony Bridge. If you stand there facing Fifth Avenue and throw down a lot of bread, you'll see them, all right. They'll all come out and push and shove and gobble up every crumb. You should come and do it, Holden. It'll make you so happy, it'll just about *kill* you.

ACT II

Moving to Fifth

My heart in hiding stirred for a bird, — the achieve of,
the mastery of the thing!

GERARD MANLEY HOPKINS

SCENE ONE

New Romance

It was the first week of May, and though tragedy had just struck Central Park's first nesting hawks, there was no cancelling the annual extravaganza about to fill the park's peaceful woodlands with wall-to-wall birdwatchers: the height of the spring migration. On May 3rd Tom Fiore and a number of other Regulars had enjoyed what birders call a Big Day. A huge wave of songbirds had arrived early that morning, and between 7:00 a.m. and 7:00 p.m., Tom and his band managed to rack up a list of 105 species of birds, including 29 varieties of warblers. "My biggest day in the park, ever!" Tom wrote in the Register, obviously unaware that he was to easily surpass that record the next year, and surpass *that* one the year after that.

May 8th, on the other hand, was the kind of day serious birders call "quiet." That's another way of saying it's a god-awful day, a day they should have stayed in bed. Only one good warbler—the hooded—was sighted that day, briefly, at Willow Rock.

Yet the day was far from a total loss. At three in the afternoon on that very day, completely recovered from his collision with a wall four days earlier, Pale Male was brought in a cardboard box to the small meadow below Belvedere Castle—the same meadow where the nectarine tree was still producing edible fruit that year. There he was released. Free again, free to wander his old haunts in the park, to range from the pond at 59th Street to the Meer at 110th, to soar from the Metropolitan Museum of Art on one side of the green rectangle directly to the American Museum of Natural History on the other (if the winds were right), to revisit his former nest sites where the memory of his great dark mate was soon just the faintest of pangs.

She was gone, but he had no urge to look for another mate just then. That season had passed, and the hormones would not begin urging him to look for love until the following December. Food was the main thing now, and there was plenty of it in Central Park. The best hunting places were precisely where his prey abounded: the Azalea Pond feeding station, for instance. A large colony of rats had come to depend on an easy food source of their own there—the birdseed, bread, dog food, and other foodstuffs scattered by year-round bird feeders. Pale Male contributed to the evolutionary future of the rat colony by availing himself regularly of those individuals less adapted to survival, or perhaps those too young to know any better. Occasionally he'd catch an avian appetizer or two—a sparrow or a titmouse and at least once, according to a horrified witness, a handsome male flicker.

SHORTLY AFTER HIS ARRIVAL in Central Park, Pale Male had discovered a hunting ground that was to become his favorite: an area near the park entrance at Fifth Avenue and 79th Street—the killing corner, as the Regulars dubbed it. A man fed a sizable flock of pigeons there every day. Easy prey. The light-colored hawk would sit on one of his two favorite

Fifth Avenue perches: the Stovepipe, a black chimney on the roof of a building on 78th Street, or a black railing at an eleventh-floor window at 80th Street. He'd wait until the pigeon flock had settled in to feed on the large quantities of grain their benefactor provided for them. Peering down intently, Pale Male would search out one that was imperceptibly slower, clumsier, stupider. Then he would plummet down in that breathtaking dive falconers call a stoop. Bingo.

Often he'd take his prey directly to a large half-dead oak in the middle of a field just east of nearby Cedar Hill: the Killing Tree. He'd land on the ground beneath or on a horizontal branch, wait for a minute or two in a posture called mantling: talons tightened around his prey, wings slightly spread to shield it from sight. He'd stand there waiting for death to come. The meal followed.

That winter, Pale Male's second in Central Park, the easily recognizable hawk was reported regularly in the Bird Register. He was now fully mature and sported a rich, russet-red tail. Though wary of anthropomorphizing, the birdwatchers found it hard to avoid thinking that he looked lonely, bereft without his mate. Then, as the year ended, Pale Male caused another flurry of excitement.

On New Year's Day Norma Collin reported a sighting of the light-colored red-tail perched on a hexagonal tower at Fifth Avenue and 76th Street. This was not an unusual place for him to be sitting—it was one of his favorite perches. This time, however, he was not alone. Another red-tailed hawk, a considerably larger one, was sitting beside him. Could it be his love from the previous spring, the one the Regulars refer to as First Love? She had been taken to Len Soucy's hawk hospital six months earlier. Whatever had happened to that bird?

❧

PHONE INTERVIEW with Len Soucy:

"First let me get my records. Oh yes, the female from Central Park. She was brought in on May 3rd last year. The bird's left wing was luxated at the elbow, at the joint where it articulates with the humerus. The ulna and the radius were luxated—that means dislocated. It was a traumatic injury; it stretched the muscles and tendons. The prognosis for such an injury is guarded.

"We taped the wing, rested the bird, and kept it in the intensive-care area of our infirmary for the better part of a month. In June the bird went outside. On June 14th it was put into a large aviary outside, but it couldn't fly. It flapped but couldn't sustain itself in flight.

"From June to September the bird's wing still drooped and we had serious reservations about this bird being viable in the wild again. We had her X-rayed again and looked at twice by our veterinarians and our staff. Again, not much improvement.

"We thought we were dealing with a cripple, and as long as we knew it was a female, we figured we could keep her here permanently and use her in our captive breeding program. I have two hundred other crippled hawks here, and one more doesn't matter a damn. I liked the thought of using it rather than destroying it.

"Then, as the fall progressed, through September and October, the bird was actually seen flying short distances. I put the bird in our flight cage, a monstrous chamber, 100 feet by 20 feet by 20 feet. You can see a bird in free flight there. And in October, unbelievably, the bird started to fly! The wing still drooped, but not as appreciably.

"By November this bird was flying the length of the cage, catching its own prey in the cage. It was a miraculous recovery—I can't explain it medically. But now that the bird was getting spicy and viable, I was afraid to keep it through an entire winter. It could trash itself in a cage, break its feathers, or further injure itself.

"I made the decision to let the bird go. But rather than return the bird to Central Park, where it had had some serious problems, I thought I'd release it in the Great Swamp National Wildlife Refuge, which is my backyard—eight thousand acres of protected territory. No hunting, no nothing. If it couldn't make it there, it couldn't make it anywhere.

"On November 7th, half a year after the bird arrived, I banded her at the tibia tarsus—that's the top of the foot—with an aluminum tarsal band that she'll wear indelibly forever—U.S. Fish and Wildlife marker 1387-38569. Then I released the bird in the Great Swamp. And she flew away like a wild bird! Unbelievably."

Though Soucy is a scientist, as he ended his story his voice was emotional.

"I have not heard from the bird since," he said after a pause. "No news is good news."

❧

ALMOST EVERYBODY had an opinion on the identity of Pale Male's new companion. Sharon Freedman, a Regular with a particular interest in hawks, voted for a new bird. Norma and Charles thought it must be the first mate back from the hawk hospital. The timing was right: Len Soucy had released her on November 7th. She could easily have headed right back to Central Park to rejoin her light-colored mate. Some hawks migrate thousands of miles each year, after all, and it was a mere 30 miles as the hawk flies from Millington, New Jersey, to Central Park.

There was one sure way to find out. Wherever she was, Pale Male's first sweetheart now wore a Fish and Wildlife Service band on her ankle. If the female currently hanging out with Pale Male had a silver-colored band on her ankle, she was likely to be the same bird. If she was unbanded, it would be definitive proof that this was a new mate.

Banded or unbanded? For the next few weeks the Regulars kept their eye out for the female's ankle. But as luck would have it, the hawks seemed to be spending most of their time looking out at the park from high Fifth Avenue rooftops. Even the most powerful binoculars could not make out a band at such a distance. It would take a closer sighting.

Finally, the female landed on a low tree near the Boathouse one afternoon in late January. She was coolly dismembering a pigeon before an excited audience of thirty or so spectators when Tom Fiore arrived, raised his binoculars, and got a perfect view of both ankles. No silvery band. This was a different female.

"I *knew* it was a new female," said Sharon Freedman, who enjoys being right and almost always is where hawks are concerned. "Her head is so much darker than last year's female." Sharon assigned her a name: Chocolate. Calling the hawks by given names, it should be said, was not an example of soppy sentimentality among the birdwatchers; it was a matter of convenience. Much easier to say Pale Male than "the light-phase red-tailed hawk." Similarly, quicker to say Chocolate than "the new female mate of the light-phased red-tailed hawk." It was around this time that the Regulars began to refer to the original female as First Love.

One morning in mid-February as Chocolate was perched on the Killing Tree, Pale Male arrived with a freshly caught white pigeon. He landed on the branch beside her, allowed her to reach over and grab his bird, and then quietly watched as she energetically devoured most of it: mate feeding—a sign that hawk romance was escalating. By the end of the month the two were observed in courtship flights, soaring and diving together. In the birdwatching community hope began to grow that there might be another nesting attempt.

On March 1st the drama of the Fifth Avenue hawks officially began. An item in the Bird Register served as a curtain raiser:

> March! Noonish, a beautiful day. Seen from Boathouse, *Red-tailed Hawk* flying south-east, carrying a woody branch. Nesting??
>
> Tom Fiore, Sarah Elliott, Marie Winn

❧

DURING THE SPRING and fall migrations Starr Saphir leads park bird walks that are noted for their extraordinary sightings. Birds that others would die for—Starr's group has two or three of them on a single walk. (Serious birders don't *see* birds, for some reason. They *have* them.) On May 10th, 1993, Starr's summary in the Bird Register (always signed with the symbol of a star) included the hard-to-find Kentucky warbler; her group had it just west of the Maintenance Meadow. Then, near the Upper Lobe, a least flycatcher, a very tricky bird to identify, darted out. Starr identified it by its faint call: Che-bek! Then the group ran across three scarlet tanagers north of the Azalea Pond. Oops, as Starr often adds in her Register entry, I forgot to mention the cerulean warbler on Cherry Hill, another dream warbler.

The rest of the year Starr pursues birds in various exotic spots on this continent and most of the others. Her ornithological skills and her not inconsequential personal gifts combine to make her a most successful walk and tour leader; her walks are always crowded, her tours generally filled to capacity, and she manages to support herself through these activities.

An actress before birding took over her life, Starr's a good-looking woman. These days, dressed in her usual uniform of snug blue jeans, sports shirt, and a canvas vest with multiple pockets, her Zeiss 10×40s around her neck and her hair in a single long braid, she looks younger than her age of fifty-something. She's an enthusiastic, highly energetic, impassioned birder; when she thinks there's a Connecticut warbler lurking in the underbrush nearby, she is like one possessed. She's also famous for her puns, swift and inexorable. One year, when she heard that Pale Male had caught an escaped parrot, she uttered one of her most famous (or infamous) quips: "That was merely an a-parrot-if." If there is such a thing as a preconscious mind, that subterranean stream of words, names, snatches of songs, associations, rhymes, and images said to influence poets, punsters, and Freudian slippers (it flows somewhere between the surface of consciousness and the depths of the unconscious), Starr's is less a stream than a rushing, overflowing torrent.

On the morning of March 2nd, Starr was leading a walk for the New York City Audubon Society. At Bow Bridge, one of the most beautiful of the park's bridges, the group paused to admire the view. Standing on the bridge and facing east, you can see the Fifth Avenue skyline; turn the other way and

Central Park West is revealed. That's when Starr and her Audubon walkers had one of their great sightings.

"We were about to cross the bridge," she related a few days later, "when suddenly I see a red-tail with a stick in his mouth flying over Fifth Avenue. It looked like a male—small. I watched it land on the northern edge of what I call the wedding-cake two-tiered building [subsequently dubbed the Octagonal Building by hawkwatchers]. It landed, and suddenly another, bigger hawk appeared right in front of him. We're all watching like, well, hawks, to see what was going to happen. He flew off his perch, still holding the stick. Then they did something quite balletic which I've seen red-tails do before. They circled around each other as in a dance. They did this for a few moments, circling and soaring, and then they both landed on a building two blocks to the south. The one they now say is the nest building."

There was a barely perceptible pause that made my ears perk up: Attention, pun coming. "It was an absolutely raptorous moment," she said, and waited another infinitesimal moment. I laughed. "If this is really a red-tail nest," she continued, "it will be the third species of hawk nesting in Manhattan. So far there's peregrine falcons and kestrels. Both of these, of course, are known to nest on buildings. A few years ago I had a kestrel nesting on a courtyard ledge of my own apartment house in Washington Heights. But red-tailed hawks . . ."

For once Starr Saphir was struck speechless.

SCENE TWO

Buteo Jamaicensis

The red-tailed hawk (*Buteo jamaicensis*) is a large, round-tailed, broad-winged bird with a wingspan of four feet, give or take a few inches. One of North America's most common birds of prey, it has an ability to soar and hover that, together with phenomenal eyesight, allows it to hunt small animals on the ground from a great distance. It sits on a high perch inspecting the world below. Suddenly it dives down. The unsuspecting creature at ground level never knows what hit it.

Carpe diem is the red-tail's motto: Seize the day! (Seize the prey is more like it.) Though designed to hunt rodents, these opportunistic hawks will eat what they can find. Rats, mice, voles, gophers, moles, shrews, chipmunks, squirrels, cottontails—these are their designated victims. But larger mammals such as jackrabbits and woodchucks are also on their hit list. Porcupines? No problem. Snakes? Yum. Frogs? Why not. Birds? Yes, mainly the ground-feeders like doves and pigeons, though Central Park birdwatchers have seen an enterprising red-tail grab a songbird in flight on several occasions. Even insects—grasshoppers, crickets, beetles, and such—are grist for a hungry red-tail's digestive mill. As the authors of *Hawks in Flight,* a useful guide to the confusing world of hawk identification, write: "Any furred, feathered, or scaled creature that is smaller than a groundhog and turns its back on a meal-minded Red-tailed Hawk might safely be said to be courting a shortcut toward the cosmic."

Farmers once slaughtered red-tailed hawks by the droves —chicken killers, they called them. True, the red-tail *will* eat chickens, but only sick, tired, old, frostbitten ones, according to A. K. Fisher, who wrote *The Food of Hawks and Owls* in the 1930s. An early conservationist, he tried to promote the sensible idea that these "varmints" provided the best form of rodent control available.

In its choice of nest site, the red-tailed hawk is similarly adaptable. It commonly builds its stick platform high in a leafy tree—an oak or elm or maple that offers a commanding view of the landscape. But if a pair of determined red-tails find themselves in treeless terrain, they'll scrounge around for alternative housing. The top of a nice cactus will do. Lacking that, they'll seek out other places with a good outlook. In 1968 one red-tail couple found themselves the ultimate of scenic nest sites: a cliff over the Niagara gorge.

But red-tailed hawks constructing a stick palace in the heart of New York City atop an ornate twelfth-floor window on one of Fifth Avenue's most exclusive apartment houses—that goes beyond the beyond of adaptable. No one had ever heard of a red-tailed hawk nesting on the façade of a city building.

WHEN TWO OF AMERICA'S greatest hawk experts were informed that a pair of red-tailed hawks were building a nest on the façade of an apartment house on Fifth Avenue and 74th Street, their response was identical: disbelief.

Dean Amadon, Lamont Curator Emeritus at the American Museum of Natural History in New York and co-author of a book on hawks so widely known it is always referred to as Brown and Amadon, was stunned by the news. "There have been reports of red-tails trying to nest in trees in towns occasionally—but never on the ledge of a building. I find it amazing," he said, adding, "That male must have a screw loose somewhere."

"Are you sure they aren't peregrine falcons?" asked Charles R. Preston, curator of ornithology at the Denver Museum of Natural History, when I called him a few days later. He was then working on the red-tailed hawk account for *Birds of North America* (*BNA*), an ambitious reference work in progress sponsored by the American Ornithologists' Union, the Philadelphia Academy of Natural Sciences, and the National Audubon Society. The man writing the red-tailed hawk account for the *BNA* would surely be considered the preeminent expert on that species.

"I don't know of any other incident where red-tails have used the side of a building for nesting," he stated flatly after I assured him that many observers, including some from the American Museum of Natural History, had confirmed the identification of the Fifth Avenue pair. "But none of it completely surprises me," he added. "Red-tailed hawks are an amazingly adaptable species, and they have been known to use various other man-made structures for nesting."

"Why do you suppose they've picked that particular spot?" I asked.

"Two things red-tails look for in a nest site are free access with an open flyway, and a good overview," he answered. "What you describe fits both of these requirements." He thought for a moment and added: "One feature of these birds I've observed that isn't often reported in the literature: when they're unsuccessful at a particular nest site one year, they'll usually move to quite a different type of site the next year, but still within the same general area. I wonder if this could be the case here."

His speculation was right on the mark. The previous year's nest in the elm at SummerStage *had* been nearby, *had* been quite a different type of site, and had, of course, failed miserably. "You could certainly argue that there's an element of learning in that behavior," he said in the cautious way scientists have of expressing themselves. I had a distinct feeling that this man thought highly of red-tailed hawks.

Our conversations with the experts demonstrated how much is unknown about birds and bird behavior. This inevitably activated a dream that many birdwatchers share: Perhaps through our observations of this remarkable pair of birds we could make a contribution to science.

An amateur adding to the storehouse of scientific knowledge is not as outlandish a possibility in the science of ornithology as it might be in physics, say, or math. As Frank Gill, vice president for science of the National Audubon Society, has said: "Ornithology has always depended on an army of amateur experts who are out there finding out things about birds."

It doesn't happen often, but it happens. When Charles Preston's *BNA* account of the red-tailed hawk was finally published, it included a notation about *our nest.* The Central Park birdwatchers had made their first contribution to science.*

---

* After an article about the nest and its ornithological uniqueness appeared in *The Wall Street Journal,* two other nests on city buildings came to light: one in Dallas, Texas, the other in Troy, New York. Yet neither of those nests, though officially within city limits, was in quite as urban a setting as Fifth Avenue and 74th Street.

SCENE THREE

Watching the Action

❧

On March 3rd the Bird Register proudly carried the first announcement:

> Red-tailed Hawk's NEST on 5th Avenue and 74th Street. Nest is on top of center window, carefully tucked behind spokes of anti-pigeon wire.
>
> Norma Collin, Charles Kennedy,
> Murray Liebman, Elliott Zichlinsky

The hawk building, as it soon came to be called, was situated smack in the heart of a well-heeled neighborhood dubbed the Gold Coast by the *American Institute of Architects Guide to New York City*. The nest itself was located on a curved ledge above the middle of three twelfth-floor windows. The birds had wedged their stick palace firmly into a grid of metal spikes installed to protect a handsome ornament from pigeon desecration—a bas-relief depicting two sad-faced cherubs holding up a medallion: the hawks, however, served as a far more effective pigeon deterrent. Meanwhile the spikes kept the sticks and twigs in place.

Directly above was an ornate, overhanging cornice, sheltering the nest and the birds from rain or snow. The nest's orientation—facing southwest—also served a purpose. When the fiercest blizzard in a century struck New York that March with winds of gale force knocking down street signs around the city and uprooting many trees in the park, the nest of the Fifth Avenue hawks emerged unscathed. This was almost certainly not a matter of luck, for several studies note that red-tail nests tend to be oriented toward the south or west.

Even the ornamentation of the hawk building seemed fateful. On each side of the façade a few feet below the nest, just above a frieze of sea serpents with tails intertwined in a fleur-de-lis pattern, stood two sculptured birds of prey. They were probably meant to be eagles—but the hawkwatchers liked to think that they *could* have been red-tailed hawks.

Best of all, the nest was in perfect binocular view. The ideal hawkwatching spot was a bench near the bronze statue of Hans Christian Andersen, a landmark especially favored by tourists who like to photograph their children and occasionally each other on the lap of the Danish storyteller. The great Dane is reading a page of "The Ugly Duckling" to an odd audience . . . a metal duck.

The bench where the watchers gathered—the hawk bench—was located on the west side of a small body of water officially called Conservatory Water. (A conservatory originally planned for its eastern shore was never built.) In spite of its fancy name, for obvious reasons this oval-shaped, concrete-rimmed pool at 72nd Street and Fifth Avenue is generally called the model-boat pond. That first year, and in years to come as the story of the Fifth Avenue red-tails continued, the hawk bench was the gathering place of the ever-growing company of those in on the secret of the red-tailed hawks.

THE MODEL-BOAT POND is not merely a tourist attraction. Many New Yorkers pass by on their way to various workplaces on Lexington or Madison avenues or to doctors' offices in zip code 10021 or to the Palm Court for tea or Rumpelmayer's for hot chocolate or a quick stop at A La Vieille Russie to pick up a Fabergé egg or two. Among them one may find Fifth Avenue grandes dames, doctors, poets, lawyers, filmmakers, artists, and celebrities' dog walkers.

Attracted by the little crowd ever watching *something* through binoculars and telescopes, intrigued, finally, by the periodic cries of excitement escaping from these enigmatic watchers—"Oh wow, did you see that landing?" or "Look, he's got a rat!"—not a few of the regular passersby ended up joining the Regulars on the hawk bench before or after work every day, or on weekends or holidays. Some stayed all day long, week after week. They became a definable entity of their own—the Fifth Avenue Hawkwatchers.

All through March an audience at the hawk bench followed the birds' every move. They watched the hawks plucking live twigs for the nest from nearby red maples and London planes. They saw them stripping bark for the nest lining from the lindens on Pilgrim Hill at the south end of the model-boat pond. One day a small group saw Pale Male pick up a scrap of old linoleum used as a makeshift sled during the last snowstorm. (Pilgrim Hill is one of the park's main sledding slopes.) "They're remodeling the kitchen," someone declared as the hawkwatchers watched the bird struggle to fly with it to the nest.

Each twig or bit of lining material required a trip from tree to nest and then back into the park again—there must have been a thousand such back-and-forth flights. The process was inspiring to witness: the birds' dedication to their task; their dogged persistence; their complete involvement in the work at hand.

The hawkwatchers' admiration was not at all diminished by the knowledge that nest-building behavior is hormonally driven. On the contrary, knowing that the increasing length of days in the spring stimulates the birds' pituitary gland to release certain chemicals and knowing that one of these triggers the gathering of twigs, and then when the nest reaches a certain stage of completion, another chemical kicks in, leading the birds to gather *different* materials for the lining—this knowledge caused many of us to reconsider our own behavior: might far more of it be triggered by instinctive messages we cannot understand or control?

ALL THROUGH THE nest-building phase the hawks carried on their sex life atop various nearby rooftop ornaments, TV antennas, balconies, and railings. Mating was almost always preceded by a gift for the female, a nice rat or freshly caught pigeon being the usual love token.

Hawk sex, like nest building, is triggered by hormones stimulated by the length of daylight. The reproductive organs of both males and females increase vastly in size during the breeding season—among some species of birds they become several hundred times larger than during the rest of the year. The enlarged testes of the male secrete greater amounts of male hormones, which stimulate the courtship behavior that precedes copulation, as well as the act itself. The female's enlarged gonads, similarly, release hormones that govern her response to courtship and her receptivity to copulation.

The sex act itself is not particularly dramatic. The male comes in for a landing from above, his talons extended in a posture the texts call the "talon-drop." He lands on her back, keeping his balance by slight wing movements. The actual hawk sex act rarely lasts more than five seconds, at least not as performed by the Fifth Avenue hawks and carefully timed by the hawkwatchers.

❧

ON MARCH 31ST the much-awaited first warbler of the spring migration—that year a yellow-rumped warbler—was sighted in the Ramble by Joe and Mary Fiore. Tom's parents—Joe is an artist and Mary a poet—are Regulars too, and occasionally the three may be seen strolling companionably together, a warming and somewhat anachronistic sight for the late twentieth century.

Finding the first spring warbler is a big event in the Central Park birdwatching community. The front-runners have been trickling in for weeks—flocks of grackles and the woodcock in mid-February, the phoebe in early March, along with the fox sparrows and the trilling juncos. The first warbler signals the real beginning: They're coming! It's happening! Soon they'll be pouring in—tanagers and grosbeaks and orioles and twenty, thirty kinds of warblers.

But a new kind of excitement overtook the gang at the hawk bench on that same afternoon. At 4:45 p.m. as nine hawkwatchers sat at the usual gathering place and watched the hawk pair at the nest, their ornithological attention was suddenly diverted from the hawk couple to a different one.

"Look! The next building!" someone exclaimed.

Nine pairs of binoculars turned to the top-floor balcony of the building just across the street from the hawk nest, the very balcony the hawks particularly favored as a trysting spot.

There had been a rumor that the duplex penthouse just below the imposing water-tower belonged to Woody Allen. Now it was confirmed. There, indeed, was the famous actor and director, live and in person, and with a recognizable companion—his youthful paramour Soon-Yi Previn. The notorious couple strolled along the terrace hand in hand while the hawkwatchers gawked shamelessly.

SCENE FOUR

The Fellowship of Hawkwatchers

Anne Shanahan's fascination with hawks began a few months after the Fifth Avenue nest was discovered. She remembers the moment well. On Friday, May 21st, at 1:15 p.m. to be exact, she was walking up 82nd Street between Madison and Fifth with her little dog Bijou, a bichon frisé. Just as she approached the Metropolitan Museum of Art, she saw a hawk swoop down to the sidewalk a mere ten or twenty feet in front of her. "It gave an incredible scream as it dove down," she recalls. "What's that?" she asked a workman outside the museum, and he answered, "It's the red-tailed hawk hunting." He said it matter-of-factly, as if a hawk hunting in the heart of Manhattan were a commonplace event for him—that impressed her particularly.

She watched the bird on the sidewalk and quickly snapped a few shots of it with a point-and-shoot camera she happened to have with her. Then she saw the hawk fly to the balcony of a small building on Fifth Avenue. It perched there for a few minutes before flying directly into the park. She knew she wanted to follow it. She entered the park too. And that was the beginning of a new life.

A slender, soft-spoken woman, neither young nor old, who listens far more than she speaks, Anne has been following the hawks in the park for more than five years now, photographing them with ever better equipment and greater artistry. She takes a great interest in the other wildlife of the park too, and her photographs are beginning to show up at exhibitions, including a recent one at the American Museum of Natural History. One of her butterfly photos appeared on the cover of *Mulberry Wing,* the magazine of the New York City Butterfly Club. But hawks are her passion.

Anne knows the Fifth Avenue hawks. She knows their favorite roosting places, their most successful hunting perches. She can detect their presence long before anyone else does. Anne rarely carries binoculars, occasionally using the long lens of her camera for a closer look at something that has caught her eye or, more often, her ear. Using a keen sense of hearing more than her eyes, she listens for sounds from other birds and especially other animals indicating the presence of a predator. Squirrels, for example, flatten themselves on tree trunks and branches and emit a low steady whine when a hawk is in the immediate vicinity. Using such clues, Anne regularly locates one or both of the red-tail pair when others have gone right by them.

Though Anne sets forth alone in the park, most of the Regulars unobtrusively attach themselves to her whenever they can. For she always seems to arrive at a place just as the hawk action is about to begin. "There's a hawk somewhere over there," she'll say as she runs into a little group of birdwatchers. Only then will the others hear what Anne's ear alone had detected: regular, sharp little chips coming out of a nearby tree—alarm notes of robins announcing a hawk in the vicinity. She can tell one boisterous blue jay cry that means "Here I come! Stay out of my territory" from a similar cry that means "Hawk nearby! Come on, birds, let's mob 'em."

Anne is modest and self-effacing, rare and devalued traits

in an era when self-esteem is the summum bonum. Yet her intimate knowledge of the hawks has gained her the admiration and respect of all—the Regulars, the Hawkwatchers, even the Big Guns. When there is a question about the hawks that requires resolution, Anne Shanahan is the first to be consulted.

Her enthusiasm is infectious, and somewhat surprising for a woman of gentle sensibilities, since most of the hawk action she follows would not receive a PG rating: Pale Male plucking a recently caught pigeon on the Killing Tree; the pair soaring together in a courtship flight, talons locked, tumbling down in the air; the couple mating on the railing of the Octagonal building—each time she witnesses such a hawk event she grows excited, as if it were the first time she was seeing it and not the hundredth. "Isn't this wonderful?" she'll exclaim, always a little breathless. "Aren't we lucky to see this?"

❧

ONE DAY EARLY in the hawks' nest-building phase, I arrived at the hawk bench and found myself the only one there. When I raised my binoculars to the nest, my feeling of isolation grew: nobody at home up there. Standing at the edge of the model-boat pond, I scanned the sky in all directions. I needed my red-tail fix.

As I stood there staring upward, an unkempt young man came up to me in a state of excitement. His somewhat sinister aspect—the impression was partly based on grooming, or lack of it, partly on some indefinable "attitude"—made me glad I hadn't run into him in a less populated part of the park. "You looking at birds? There's a humongous bird on a tree over there. It's eating a pigeon!"

He half walked, half ran in the direction of Pilgrim Hill. I followed him, and a few minutes later I was not surprised to see that the humongous bird was Pale Male. Perched on a low branch of an evergreen halfway up the hill, he was busily eviscerating a plump dark-gray pigeon. Feathers were flying. The hawk's talons and beak were visibly bloody, and something about the perspective of our vantage point made him look particularly huge.

The comradeship of hawkwatching swiftly dispelled my previous caution. In the way of the hawkwatchers, I introduced myself by first name to the young man. He did the same: Johnny. Together we stood there for a good ten minutes, transfixed.

Suddenly, as if from nowhere, another hawk materialized on the branch beside the dining male. It was the female, the one we had started calling Chocolate.

"Holy shit!" said Johnny. For the next ten minutes he and I and the lady hawk watched Pale Male scarf down the pigeon. All three of us gazed with unwavering attention.

When the pigeon was almost consumed, the young man and I heard the female give a little snort that I translated to mean: "Some gentleman you are!" A moment later she reached over with a talon and grabbed the remains of the pigeon. Pale Male tugged for a few seconds, but without much enthusiasm. Then he emitted a snort of his own that I understood as: "Hey, I was just about to pass it over." With that the pale hawk flew off the branch and headed for Fifth Avenue, while his lady finished the pigeon.

I lent Johnny my binoculars and pointed out the nest. I knew exactly the moment he found the nest because he uttered the same comment as before. He was genuinely flabbergasted to learn that hawks had taken up residence in such an unlikely spot.

The next day Johnny reappeared at the hawk bench. He

had brought a friend—José—to see the hawks. This time, as luck would have it, both hawks were sitting on the nest. I passed Johnny my binoculars and he was just as enthusiastic as on the previous visit—a potential birdwatcher, I thought. José, it must be said, seemed to be in a different space entirely.

Part of hawkwatcher etiquette is not to ask the usual cocktail-party questions, "What do you do?" and the like. But curiosity is always there, so I was glad when Johnny's next remark answered some of my unspoken questions about his life history.

"I told my caseworker at drug rehab that I saw a pair of hawks making a nest on a building on Fifth Avenue," Johnny told me. "She looked at me funny and asked, 'What have you been smoking?' "

SCENE FIVE

High Anxiety

One day in early March, while the nest was still under construction, workmen appeared on the roof of the hawk building, causing general alarm, if not panic, among the hawkwatchers. Wouldn't human activity so near the nest cause the hawks to abandon it? Hadn't Arthur Cleveland Bent declared: "A brush blind is utterly useless [for observing a red-tail nest] as the hawks can see the slightest movement in it, and will not come near the nest again until the intruder departs."

Even when a workman appeared on a rooftop several blocks away from the nest building, the hawkwatchers worried. Hawks, we knew, have vision far beyond human abilities. According to Roger Pasquier's *Watching Birds: An Introduction to Ornithology,* the small retinal bodies called cones that allow the eye to form sharp images and distinguish shades of color may be as dense as 1 million per square millimeter among hawks, while man has only a fifth as many in the same area. Eyes like a hawk, as the saying goes.

Any window washer in the vicinity was cause for alarm, and there were many of them—it was spring-cleaning season and this was the elegant East Side. When a scaffold appeared on the hawk building itself one day in March, it was a five-alarm emergency.

"Let's go over and see what we can do," Charles Kennedy proposed. And we did.

Charles, an established Regular, is a tall, lean, good-looking man with a slightly countercultural look. He is a good birder and an excellent forager who together with Norma Collin delights in judiciously sampling the park's edible bounties. His tastes are more eclectic than Norma's, however, for not only does he favor black cherries, juneberries, mulberries, blackberries, wineberries, and other reasonably well known edibles, but he alone considers fresh elm seeds a tasty snack. He's the only person I know who considers trees a form of life equal in value to humans. Yet his human skills are not inconsiderable: everybody likes Charles. I knew he'd be a tactful ambassador to the powers-that-be at the hawk building.

We rushed across the street. There, on the 74th Street side, two workmen were just bringing the scaffold down to ground level near the service entrance. Charles smiled at them, courteously introduced himself and me to them, and shook hands with each of the two. They stared at him as I imagine the villagers near La Mancha might have gazed at Don Quixote when he arrived in their midst to battle with windmills. As Charles explained at some length the history of the red-tailed hawks and provided them with salient information about the species *Buteo jamaicensis,* I noted that the workers' jaws literally dropped as they waited for him to conclude his lecture. I grew a little uneasy—they were staring at him so intently, so fixedly. But it soon became clear that they were merely trying to eke out a bit of meaning from what he was saying. One of them spoke no English at all. The other—his name was Rahaman, he told Charles—spoke a little.

Charles gently questioned Rahaman with words and gestures and learned that they had finished their repair job on the roof cornice—they weren't window cleaners after all: We didn't have to beg them to stop anything. Rahaman finally got the drift of our inquiry when Charles spread his arms and flapped them slowly and mightily. "Yes yes yes," he said with a flash of understanding at last. "Eagles up on roof, big eagles."

The hawks resumed bringing twigs to the nest as soon as the scaffold was gone. These were not shy country hawks, after all. They were urban birds, Big Apple hawks who paid no notice to workmen, window washers, or, indeed, on so many occasions in years to come, to park spectators who sometimes came so close they could almost touch a feather.

❧

AS THE WEIRD REALITY of the Fifth Avenue hawk nest sank in, we began our efforts to reach the superintendent of the hawk building. What if something happened to a future baby, we worried, what if it fell out of the nest. We'd need the super for access to the roof, from which, we figured, someone could pop the kid back in.

The simple matter of the super's name was the first stumbling block. For just as Cerberus guards the gates of Hell, so at the entrance of each elegant building on the western border of zip code 10021 stands the uniformed Doorman. This implacable watchdog is there to check the visas and passports, so to speak, of anyone darkening that canopied Door of which he is the Man. He—for there are no doorladies on Fifth Avenue—adroitly fields questions like "Can you tell me the name of the super of this building?" with polite indirection: "Do you have a package you wish to deliver?" And if, God forbid, you're not merely asking questions but seeking to make your way *inside*—as Charles and I actually did a few years later for a hawk-related purpose—in such a case, the doorman has various ways of making you feel like you're Jack or Jacqueline the Ripper in disguise.

There are good reasons for this passion for privacy on the Gold Coast: a great number of the super-rich and super-famous live there. We knew of one super-superstar in the hawk building itself—Mary Tyler Moore. She had a floor-through four stories below the nest. And there was Woody across the street. With assassins, kidnappers, burglars, extortionists, mad bombers, and terrorists making headlines every day, and a legion of paparazzi ever on the lookout for prey, it is clear that employees of Gold Coast buildings must be exceptionally careful.

Fortunately, Annabella Canarella, a flight attendant and one of the keenest of the Regulars when in town (she does the Tokyo run every few weeks), turned out to have a friend who was a friend of Mary Tyler Moore's dog walker, Jody. Jody provided the super's name: Hugo Navarette. His office was in the building's basement. And he wasn't even called a super, we discovered. On Fifth Avenue the job is called building manager.

THE HAWKWATCHERS NEEDED to conduct some delicate negotiations with Hugo. First we had to make sure he was aware of the nest's very existence. What if somebody mistook the pile of sticks above the twelfth-floor window for wind-borne trash? We worried about the people in the hawk apartment—what if they decided to tidy up the ornament above their middle front window? The idea that someone might destroy the nest in the name of cleanliness was too terrible to contemplate.

How well we knew, from many sightings, that these were not simply sticks gathered from the ground. Each twig had been broken off a living tree by dint of extreme effort. The hawk, as we observed again and again, would land on a tree, often awkwardly balancing on a thin branch that could barely support such weight (anywhere between $2\frac{4}{5}$ and $3\frac{1}{5}$ pounds). Using the wings to achieve balance, he or she would bite down on a branch or twig, working away at it until it came completely loose, then finally bear it away in beak or talons to the nest.

Nor was the trip with a twig an altogether straightforward matter. First the arriving hawk would circle the nest several times with the twig, sometimes flying a city block or two south or north, before finally heading directly for the nest. Maybe ancient instinct programmed this indirect approach, to avoid giving away the location of a nest. If so, it was obviously an exercise in futility since just about every move those birds made was monitored by a large audience, indeed, magnified seven, ten, or (with a spotting scope) sixty times. Maybe the hawks just needed the exercise, the watchers suggested.

On the day I first visited Hugo in the basement of the hawk building, two empty wooden crates sitting outside the service entrance attracted my attention. I looked them over for furniture potential—a conditioned reflex from college days—and noticed a label: Veuve Clicquot Ponsardin, a fine French champagne. They buy it by the case on the Gold Coast.

John, the building's handyman, ushered me into Hugo's large, neat, freshly painted office, where the super was sitting at a desk. A man of few words (understandably—his job depended on his discretion), Hugo told me he had been the building's super for over twenty years. He also revealed that he was well aware of the hawk nest on the top floor. I tried to get him to tell me who lived in the hawk apartment—everyone was dying to know—but he wouldn't give me the smallest hint except to say that the people who lived there also knew about the hawks.

When he was growing up on the outskirts of Santiago, Chile, Hugo did tell me, he and the other neighborhood kids were often waiting for some chicken eggs or duck eggs to hatch. It worked like a clock, he said—twenty-one days and they would hatch. Now he would wait and see what happened with the nest above the twelfth floor.

I didn't think someone who had once waited for eggs to hatch in Santiago, Chile, was about to trash a hawk nest. My anxiety level briefly subsided.

❧

WE HAD HIGH HOPES about Woody and Mary (as we presumptuously referred to them). Surely their fame and power could help ensure the safety of the nest. But reaching them seemed an impossible task, stars being what they are both in the heavens and on earth: inaccessible.

Again we resorted to Jody the dog walker, who obligingly left several letters from us for Mary. We gave her details about the hawks nesting four floors above her apartment, including an abbreviated history of Pale Male's first nesting attempt the year before. We sent along information about the red-tailed hawk's breeding cycle. To our amazement, Miss Moore wrote back. She included her telephone number and suggested that one of us call her. I was assigned the job.

It was thrilling to hear the unmistakable voice of Mary Tyler Moore coming out of my telephone, almost as exciting as sauntering in the Ramble and coming upon a prothonotary warbler (another legendary star). She promised to do anything she could if there were problems with the hawk nest.

Mary Tyler Moore continued to be a comforting background presence and a source of reassurance for all the years the Fifth Avenue hawks occupied the hawkwatchers' attention. It wasn't that she actually involved herself in our affairs. It was simply the idea of her. Now we knew someone in the hawk building, an important someone. We were confident she *would* step in to help if there was trouble. But in spite of our repeated invitations, she never came to the hawk bench, at least not as far as we know. Perhaps she came in disguise.

(A few years later we had further evidence of Mary Tyler Moore's commitment to the Fifth Avenue hawks. When the drama took one of its most dramatic turns, she sent a generous contribution to Len Soucy's rehabilitation center, the Raptor Trust.)

Woody Allen was the other first-magnitude star in our hawk constellation. We knew he was a self-avowed nature hater ("I am two with Nature," he has written), but he turned out to be a hawkwatcher too. I found this out from the general contractor of a renovation project on Woody's roof garden. The same man had done a job for my sister earlier that year and had casually mentioned to her his next employer—only two degrees of separation out of the proverbial six.

Naturally, I called the contractor. After I promised not to publish his name, he revealed that Woody had been alerted to the hawks' presence by a noticeable absence: no more pigeons wreaking havoc on his garden. This desirable solution to a chronic problem appealed to the actor-writer-director and he began to investigate the idea of constructing a hawk nesting platform on his own terrace, perhaps for peregrine falcons.

He never built it—why bother, if nature does the job for you right across the street, and at no expense to you. Maybe he is one with nature at last.

❧

IT WAS RAINING all morning and into the afternoon on April 1st, the day Tom Fiore recorded his second annual April Fools' Day sighting in the Bird Register: a blue-gray tanager. This tropical bird may be found in only one Central Park location: the Rain Forest Hall at the zoo. In years to come Tom would describe sightings of a keel-billed toucan and two sunbitterns skulking through the undergrowth at the southwest corner of Central Park. That, of course, is where the zoo is located, so in point of fact Tom always managed to maintain his record of strict accuracy.

It cleared by early afternoon, and Norma Collin spent from 1:30 to 5:30 at the hawk bench. She was the first to notice a distinct change in behavior of the nesting red-tails. Now the birds began to exchange places on the nest; as soon as one arrived, the other flew off. There was always a bird hunkered down on the nest from this day on.

Our interest in the hawks was no longer focused on architectural achievement or aviational accomplishment. It was a family matter now: They were obviously sitting on eggs.

Though anxieties that the hawks would abandon the nest now faded—according to Bent, hawks are far less likely to leave a nest once eggs are laid—new and more troubling anxieties took their place. Signs bearing a skull and crossbones superimposed on a line drawing of a large rat suddenly appeared on tree trunks throughout the park. They announced in English and Spanish that Quintox, a rat poison (*venenos de ratas*), had been placed in rat holes throughout the area. Panic! What if a hawk were to eat a Quintoxed rat? If the poison was strong enough to kill a rat, wouldn't it harm the animal eating that rat? What if they fed a poisoned rat to the future babies?

The odds that a hawk would eat a poisoned rat seemed high, for according to Pasquier's text: "Raptors, like other predators, usually take the weakest, sickly, least alert members of their prey species." A poisoned rat would surely fit that bill. Only Sarah Elliott came up with a comforting idea: "Maybe a poisoned rat tastes bad, or at least funny, and the hawks won't eat it."

Calls to various environmental agencies did nothing to allay anxieties about the rat poison. According to Bill Erickson, a biologist at the registration division of the Environmental Protection Agency, Quintox was mainly intended for use in and around structures where rats abound—but not in areas where wildlife might have access to poisoned rats. It had never been acceptably tested for "secondary effects" on wildlife, that is, for its impact on creatures that prey on rats and might have access to a poisoned one.

Rats, of course, have been a notorious scourge throughout human history, especially as carriers of bubonic plague. In the course of gathering information about the rat poison, one basic question kept coming up: What diseases do Norway rats in our vicinity carry that might affect humans? An eminent infectious medicine specialist at Cornell Medical School, Dr. Mark Stoekle, gave an unexpected answer: "I don't know of any specific diseases associated with Norway rats in the United States," said Dr. Stoekle, adding, "Here, rat control is primarily for aesthetics."

SCENE SIX

Lunacy at the Reservoir

It was early April. With Pale Male and Chocolate sitting on eggs at the Fifth Avenue nest, making exchanges and sitting again, we needed a break from the intensities of hawk-watching. We found it at the Reservoir.

TOM FIORE'S DAILY rounds of the park always include the Reservoir. Great numbers of gulls, an occasional heron or egret, pied-billed grebes and other waterfowl, and a surprising variety of ducks use the Reservoir as a stopping-off spot during migration, or as a regular wintering home. Always a chance of something extraordinary there, Tom believed.

Occupying some 106 acres north of the 86th Street transverse, the Reservoir is by far the largest of the park's eight water bodies. While the others, such as the Lake, the Loch, the Gill, and the Meer, are purely aesthetic creations, the Reservoir long served as a functioning part of the New York City water supply system. A chain-link fence around its 1.58-mile perimeter, mandated by the Department of Environmental Protection, was erected years ago to keep neighbors—whether human or canine—from taking a plunge in its pristine depths. From the birdwatchers' point of view, the fence served to protect not only the purity of the water but the birds that take refuge there as well.

The year had started off with a bang when an uncommon visitor to the park called, paradoxically, a common goldeneye had arrived on January 7th. A striking green-headed duck with a conspicuous white spot on the side of its face and a yellow eye, the goldeneye settled down at the Reservoir for a long stay. In March two pairs of ring-necked ducks showed up briefly. Not as uncommon as the goldeneye, these appear infrequently enough to be specially noted in the Bird Register. In addition, there were goodly numbers of the expected winter species: canvasbacks, ruddy ducks, shovelers, scaups, and several pairs of charming little black-and-white diving ducks called buffleheads. According to the great duck authority John C. Phillips, the coastal Indians used to call these creatures "spirit ducks" for their uncanny ability to appear at the water's surface as if from nowhere, ghostly apparitions.

When Tom arrived at the Reservoir on the morning of April 9th, the early fog had not quite dissipated. Perhaps that's why he thought for a moment that the bird he saw swimming and diving about two hundred yards from the south side of the Reservoir was a double-crested cormorant, a regular Reservoir visitor.

Then Tom noticed the bird's thick black bill, held fairly level with the water. Cormorants have yellow bills, usually held at an upward slant. As the fog lifted, Tom could see that this bird had a black-and-white-checkered back, as well as a lovely black-and-white necklace of markings around its neck. It was a loon!

The common loon *(Gavia immer)* is a large, powerful water bird known for its demented call. Just a year earlier, much to the delight of the park's birdwatchers, a number of these birds had spent some days at the Reservoir, the first visit of their species to the park in many decades. Of course that could have been a fluke. Now here was a loon again. A marvelous thought popped into Tom's head: since migrating loons normally stop in the same places along the way, year after year, perhaps Central Park would become one of their regular stopover spots.

During the second and third weeks of April more loons flew in, until there was a total of ten noted on April 15th. And while loons take on a drab, grayish color during the winter months, all ten birds were seen that day sporting their flashy black-and-white breeding plumage—a thrilling sight.

A little after sunrise a few days later, Tom entered the park through the Hunter's Gate at West 81st Street. (The twenty-two main entrances to the park all bear names of occupation groups: merchants, artists, scholars, and the like.) That's when he heard the wild and eerie sound of loon laughter—an unforgettable moment. Toward the end of April the loon numbers began to diminish; by May 9th all had departed for their breeding grounds, remote spruce-shaded lakes in northern New England and Canada.

"COMMON" IN NAME only (like the goldeneye), the common loon is on the "endangered" list in at least one state, "threatened" in several others, and listed as "a species of special concern" in yet others, including New York State. Wintering loons, who brave the elements at sea up and down both coasts of North America, are vulnerable to oil spilled by tankers. Acid rain, shoreline development, and disturbance by motorboats on the once-solitary northern lakes where loons breed have diminished their reproductive success. To add to their troubles, these distinctive, deep-diving birds ingest quantities of lead fishing sinkers that litter lake bottoms; the result: a high loon mortality rate from lead poisoning.

It was not only the loon that was in danger of extinction that year. The city's Department of Environmental Protection had declared the Reservoir obsolete, and a decision had been reached to stop pumping drinking water from it. Its very future as a water body was now on the line, as plans were considered for filling it in and converting it into a playing field.

There was a precedent for such a move. In 1862 a preexisting reservoir was rendered obsolete by the construction of a new and bigger one just to its north—the present Reservoir. After decades of controversy, in 1936 the old Reservoir was finally filled in and transformed into the Great Lawn.

At a public meeting called to discuss the future of the present Reservoir, the fill-it-in-for-ballfields faction made its argument first. Vigorous objections to this plan were then voiced by two powerful groups: the fortunate and often influential owners of apartments with windows looking out over Central Park who made it clear that their happiness depends on seeing that large body of water intact; and the legions of joggers who regularly pound around the Reservoir's circumferential track. They declared, every cardiovascularly fit man and woman of them, that they must have water in their peripheral vision as they huff and puff.

The visit of loons to the Reservoir inspired Central Park's birdwatchers to come up with another alternative for the future of the park's largest pond: turn it into a bird sanctuary. Retain a fence of some sort to keep out stray dogs or human trespassers, but remove the riprap at the shoreline and put in plants that would provide food and cover for songbirds as well as waterfowl.

• • •

THOUGH THE PRESERVATION of a stopover point in Central Park for a small number of loons would not have a meaningful effect on the overall loon population, indirectly it might serve their cause. The sight of a loon making a long, skittering landing on the Reservoir, or the unique sound of its crazy, wonderful call over Manhattan Island might inspire some of those powerful and influential people with windows looking out on the park to do something to further the loon cause. They could make a major contribution to the North American Loon Fund, for example, an organization that works to promote the preservation of loons. The group had already succeeded in persuading the federal government to ban the manufacture and distribution of lead fishing sinkers throughout the United States. There was still a lot to be done.

Can the loons in turn help save the Reservoir as an undisturbed refuge for wildlife? Dr. Judith McIntyre, professor of biology at Utica College and author of *The Common Loon: Spirit of Northern Lakes,* believes that the presence of loons at the Central Park Reservoir is enough to justify its continuing undisturbed existence.

"Loons in the heart of New York City? How amazing!" she exclaimed when she heard about the loons' visit to the Reservoir. "People go hundreds of miles into the wilderness to see and hear these beautiful birds. If the loons are willing to stop during migration and share their calls and beauty with the citizens of New York City, surely those citizens can maintain their Reservoir as a refuge for the loons." The citizens seemed to agree: the following October commissioners of three city agencies met at the Reservoir and signed a Memorandum of Understanding. It declared that the Reservoir would be preserved as "a permanent water feature of Central Park."

SCENE SEVEN

In Denial

❧

On the hawk bench, everyone's mood brightened as the month of May approached. Incubation had begun on April 1st—that was the first day we saw the pair make deliberate exchanges on the nest. Twenty-eight days had gone by. Based on the known incubation period for red-tailed hawks—28 to 32 days—Hatch Day was imminent.

In late April and early May Central Park birdwatching reaches its climax—the famous spring migration. That year, however, instead of the annual quest for the cerulean warbler (seen on April 30th just south of the Ramble Shed's rest rooms by Starr Saphir) and the Philadelphia vireo (seen by Sarah Elliott on May 10th on the west side of the Azalea Pond), the obsessed band of hawk addicts trained their binoculars on the Fifth Avenue nest from dawn to dusk, searching for signs that the eggs had hatched.

Seventy-five species of birds were reported in the Ramble on May 6th by Tom Fiore, who managed to remain a dedicated birdwatcher as well as an ardent hawkwatcher. By then most of the hawk bench gang had their eye on one species only—*Buteo jamaicensis,* the red-tailed hawk.

How would we know *it* had happened? Though we

couldn't see into the depths of the nest, and though we knew the newly hatched chicks would be too small to be seen over the nest rim for quite a few days, we understood that the parents' behavior at the nest would change conspicuously when the eggs had hatched. We would see the hawk parents bringing prey to the nest, then making up-and-down motions as they deposited little chunks of food into the gaping beaks of the unseen chicks deep within.

On April 29th Norma observed fresh greenery in the nest. Wasn't this said to be a sign of imminent hatching? Sharon Freedman saw a carcass being taken out of the nest on May 2nd. Didn't this indicate the parents were now feeding young? Tony Luscombe, who once supervised a hawk observatory in Peru, noted the female standing on the nest on May 6th. "Stood as if shading nest. Common behavior on a hot day with young," he wrote in the Register.

"We have a hawklet!!!" Gene Nieves wrote in the Bird Book on May 7th. One of the most fervent of the new hawkwatchers, Gene was planning to release laboratory mice on Pilgrim Hill once the chicks hatched. The parents would need extra food for their young, he declared. I hope he didn't run out and buy the mice that day, for it proved to be another false alarm.

The eggs were long overdue to hatch, and still the hawks continued to sit . . . and sit. By the middle of May optimism began to fade: The eggs were not hatching. Perhaps there were no eggs at all in the nest. Perhaps the eggs were infertile. Then again, maybe our calculations of when incubation began were wrong. It takes a long time for hope to die completely.

❦

I HAD NOT given up hope entirely when I paid a visit to Stanley Diamond, a lawyer who lives on the eleventh floor of a building on Central Park West and 74th Street—directly across the park from the hawk building. Though his apartment was considerably farther from the red-tail nest than the hawk bench was, he had managed to bridge the gap—and then some—by means of technology: in his living room window stood an astronomical telescope pointing directly into the hawk nest. While our strongest spotting scopes magnified sixty times, his instrument was more than four times as powerful as that. I cadged an invitation to pay him a visit—I knew a friend of a friend of a colleague of his wife's. A look through his telescope might end our speculation once and for all. Maybe I would see chicks at last, or at least eggs.

It was a fabulous apartment, but as soon as I walked into the living room and saw the telescope in the window, I headed for it like a horse with blinders. I dutifully wrote down its name, for I knew the birdwatchers, ever interested in optics, would ask. It was a TeleVue Optics telescope with a Nagler lens and a 2.5 doubler. I stood on tiptoe without waiting to have the tripod adjusted to my height, too eager to take the look I had been anticipating so keenly.

The image was oddly dreamlike—everything was floating as if it were on water. At first I didn't recognize the hawk. From the hawk bench, I was used to seeing just the top of the head of the bird in the nest; now almost the entire bird was visible. It was Pale Male, though it took me a while to be sure. He looked so different at that magnification—so much more, well, personal. The markings around his beak gave his face an *expression,* as if it were a human face. He looked proud, somewhat disdainful. That was my impression as I gazed through the telescope at him.

From the hawk bench, the bird in the nest looked motionless. Now I could see Pale Male's head making constant little movements—he was watchful, alert. I could see his eye clearly, and a startling illusion made it seem as if he were looking directly at me. Alas, I could see immediately that even if the bird flew off, the inside of the nest would not be visible. The angle was wrong. The window was not quite high enough to look down *into* the nest.

The only piece of substantially new information I picked up on my expedition across town was about Stanley Diamond's next-door neighbor there on the eleventh floor. It was Mia Farrow, Woody Allen's ex. Her apartment also looked out at the hawks' nest—and at Woody's building too, come to think of it. I wondered if she had a telescope in *her* window.

❧

ONE LATE May afternoon, on the seventeenth floor of a high-rise one block north of the hawk building—the Green Shade Building as we called it—a young artist named Ed Raph took a break from his work to look out the window. The window offered a view of Central Park and its great expanse of woodlands, meadows, and shining lakes surrounded on all sides by the razzle-dazzle of the New York skyline. The contrast never failed to excite him.

The apartment was not his, nor was he working on his art that day. In spite of a master's degree in fine arts from Yale, he had taken a job as a carpenter. Together with a crew of three others, he was renovating a multimillion-dollar apartment.

The window was opened wide to let out dust and plaster. As the artist-carpenter glanced out on this particular day, something happened that took his breath away: a huge bird suddenly appeared, as if from nowhere, and landed on a balcony railing a short distance below the spot where he was standing. In its talons the bird held a sizable gray-and-white pigeon. It was not clear whether the pigeon was entirely dead on arrival. But several seconds later all doubt on that matter vanished as the great creature tore into its prey.

As Ed Raph stared at the scene below, he had a sudden realization that the bird was just as aware of his presence as he was of the bird's. It paused, cocked its head, and looked up at him for a long moment before resuming its meal. If perception leaves physical traces on the perceiver's brain cells, as neurobiologists tell us it does, Ed Raph was now part of that hawk's brain as much as it was part of his.

"Come quick!" he called to the others on the crew, and soon there were five of them standing at the open window, gazing at the extraordinary sight of a large bird of prey chowing down on a plump pigeon. Richard Baronio, the general contractor of the job, raced back to his briefcase and returned with a point-and-shoot camera. He usually uses it for photographing exposed plumbing pipes before they are sealed over with plaster. This time he leaned out the window and took pictures of a hawk and its prey on a high balcony railing. Baronio happens to be an artist too, with an M.F.A. from Brown University. In fact, everyone on the crew was an artist. They enjoyed each other's company, and were planning a group show of their artwork in the near future.

Suddenly the bird paused and made a loud noise, according to Raph. It seemed to be a signal of some sort, for a moment later a second, even larger bird landed silently on the railing just next to the first one. "Oh my God, another one!" one of the workmen said. For fifteen minutes the newly arrived bird sat and watched the first bird eat the pigeon chunk by chunk, while the humans watched too, a bit squeamishly, they later admitted.

After waiting patiently for all that time, the bigger bird suddenly made a move. It hopped closer to the feasting bird, reached over with its right talon, and grabbed the pigeon. The first bird struggled halfheartedly. It gave up quickly and then sat there for another five minutes watching the bigger bird dig in. Finally, the original bird took off. The men at the window watched as it effortlessly soared and circled in front of them before flying purposefully in a southerly direction, straight down Fifth Avenue. It seemed to be heading for another building. "It was like something from *Nova,*" one of them said afterwards.

LOOKING WITH BINOCULARS from the hawk bench that day—keeping my vigil though the hawk eggs were long overdue to hatch—I had seen Pale Male arrive on the balcony railing of the Green Shade Building. Don't they ever raise those green shades on the top floors, I remember idly thinking when I noticed a workman standing at the open window of an apartment directly above the bird. He was pointing a camera at the scene below. "What a picture that would be!" I thought, and quickly biked over to the Green Shade Building, where I inveigled the uncharacteristically friendly doorman to let me go up. By the time I entered the apartment and reached the window, the female hawk had arrived and they were both on the railing.

I stared at the pair. It was the closest view of the Fifth Avenue hawks I had ever had before—or have had since. But I couldn't enjoy the beautiful sight. A terrible realization struck me: both birds were off the nest. Obviously they too were finally giving up hope. I felt a familiar pang—the pain of loss—and then I swiftly rejected the reality that was staring me in the face. No. It was an unusually warm day, that's all. They were just taking a break while the sun kept the eggs from cooling.

By the time I got back to the hawk bench, one of the hawks was, indeed, back on the nest, and everything was back to normal. But I had given the workmen my name and address, and now I have Mr. Baronio's close-up photograph of Pale Male and his mate on the balcony railing of the Green Shade Building as a souvenir of the hawks' first season on Fifth, of my meeting with the artist-workmen, and also, I suppose, of being most undeniably in denial.

SCENE EIGHT

Better Than Being in Love

The hawks sat for the expected month of incubation, and they sat an entire month longer. Of course we knew this was not really faithfulness or hopefulness on their part—just a matter of hormones. Still, we found the birds' devotion to their ill-fated nest heartbreaking.

Finally, the hawks wised up. During the first week of June (shortly after their visit to the artist-carpenters in the Green Shade Building) Pale Male and his mate ceased their constant sitting. They began spending hours at a time perching on windowsills and balcony railings of nearby apartments, sinisterly looking in for hours at a time. Sometimes they flapped their wings against the windowpanes. Imagine glancing out your window and seeing a great bird battering to get in! By the end of the week, the hawks stopped appearing at the nest altogether.

On June 13th after much prodding by the hawkwatchers Hugo the super and John the handyman went up to the roof to check out the deserted nest. Lying flat on his stomach, John leaned way out over the parapet while Hugo kept a firm grip on his feet. "Three beautiful, perfect eggs" was the report. So. The hawks had not been sitting on *nothing*.

I desperately wanted those eggs. Only by sending them to a wildlife pathologist for chemical analysis would we ever find out if there were traces of rat poison, Quintox or Ditrac (a second poison the park had begun using in May).

But Hugo the super wouldn't give up the eggs without an okay from the managing agent. "I thought *you* were the managing agent," I said, but no, Hugo was the building manager. Alvin Traub, an employee of a large real estate company, was the managing agent.

Alvin Traub wasn't quite in charge either. "I'll need permission from the board of directors of the co-op," he said when I called with my request. A few days later the verdict came down: permission to retrieve eggs denied, for "security reasons." I must have sounded pretty glum, for he relented somewhat. "They're putting up a scaffold for façade cleaning and roof repairs later in the summer. I'll tell Hugo to get the eggs for you then," he said. *If* there were any eggs left to retrieve, I thought bitterly, for surely a crow or gull or blue jay would have made off with them by then.

I kept a lookout, and when a scaffold appeared on the front of the hawk building in early July, I paid another visit to Hugo. "Please, please, get us those eggs," I implored. "Otherwise we'll never know what went wrong." I added that I had spoken to Alvin Traub, who had said I could have the eggs when work on the façade commenced. Those were the magic words.

"If the managing agent said OK, it's OK," said Hugo, and he promised he'd get me the eggs.

A few weeks later the scaffold had reached the nest. The next morning I rang Hugo's basement doorbell. Had they removed the eggs as they had promised? I asked. I was certain they hadn't. But to my surprise he answered "Yes," in his usual laconic way. Were they intact? "No," he said in the same tone of voice. I sighed and began to leave. "Wait a minute," he said and buzzed for John the handyman.

While we waited, Hugo lost some of his reserve. He began to talk about the nest and threw in a detail he hadn't mentioned before. When John the handyman had looked down into the nest that day in June, he had noticed that the eggs were partially resting on the anti-pigeon spikes. "Maybe the spikes hurt the eggs and that's why they didn't hatch," he suggested.

The handyman arrived ten minutes later carrying a huge, bulging black garbage bag. I took a peek inside and was shocked by what I saw. I wasn't getting just the broken eggs. This was the entire nest! I could smell a strong, musty odor coming from the bag.

I managed, with considerable difficulty, to get the thing back to the park, balancing it on the milk crate attached to the back of my bike. I felt terrible. The hawks' nest, our wonderful nest, in this hideous plastic bag.

I took the bag into the woods at the top of the path from the Boathouse. There, my heart beating as fast as if I'd run a mile, I spilled the contents onto the ground. The dusty, moldy smell assailed me again. If there were any airborne diseases to inhale, I inhaled them.

Visible immediately—one egg. A hole at the top, but surprisingly whole. Inside it I could see some wet-looking bluish green stuff. I wrapped it in my handkerchief and put it in a cardboard box I had optimistically grabbed from home for just that purpose. The egg smelled terrible, rotten—like an unblown Easter egg that falls and breaks six months later.

Then I sat down and began to inspect the crazy collection of sticks, downy feathers, bark, and egg fragments that had once been the hawk nest. Even the piece of linoleum was there! Those thousands of trips back and forth, back and forth—

what futility! I felt oppressed and oddly guilty, as if somehow *I* had been responsible for the nest's sorry fate, an irrational feeling that persisted for the rest of the day and even longer.

A few days later I brought Charles to see the former stick palace. He immediately dropped to his knees, as if to worship. In fact, he was carefully examining the twigs. Then he pointed out something I hadn't noticed. Every twig displayed the beak mark of a hawk. It had been bitten off at a diagonal. Charles proceeded to measure. Using his open hand as a benchmark—eight inches, he told me—he checked out about twenty sticks and found they were each between eight and fourteen inches long.

I sent the large egg fragment with the bluish-green gook inside to Ward Stone, a wildlife pathologist for the Department of Environmental Conservation in Albany. Almost a year later I received his report:

> There was no indication of embryonic development having occurred in the red-tailed hawk egg. A chlorinated hydrocarbon screen shows several chlorinated hydrocarbon pesticides are present, but not at levels that would kill the embryo. Although the Quintox and Ditrac applications were correlated with the hawk nesting attempts, they probably did not play a role in the lack of development of the eggs.

❧

IT WAS OVER. The hawks had dispersed, and so had the hawkwatchers. Soon it felt as if it had never happened. But there was an unfinished feeling in the air. We needed a real ending.

The Hawk Farewell party was Charles' idea. A June date

was set, a cake decorated with a red-tailed hawk on the icing was ordered, a telephone chain was organized. Fifteen of the loyal band attended and toasted Pale Male and Chocolate with champagne. It was a Portuguese brand, better suited to drinking from paper cups than Veuve Clicquot might have been. (Sharon and Norma, non-drinkers, made their toasts with Martinelli's sparkling cider.) Several corks popped into the model-boat pond, mildly frightening the family of seven mallard ducklings that had hatched there in early May.

Everyone reminisced about their favorite hawk story: the day after the snowstorm when Pale Male ate a woodcock; the rat feasts on Woody's terrace; the time the female caught a pigeon in midair just above the model-boat pond.

It was hard to recapture the crazy excitement, the anxiety, the obsessive interest we had all felt. "It was like being in love," one of the Regulars said, and everyone agreed.

Harold Perloff, a British birdwatcher who had been part of our band since January, had the last word. "It was *better* than being in love," he said with considerable conviction.

Intermission

"Hope" is the thing with feathers—
That perches in the soul—
And sings the tune without the words—
And never stops—at all—

EMILY DICKINSON

SPRING TO SUMMER

Baby Magic

The following year the red-tails rebuilt the nest, stick by stick. The hawkwatchers watched them build. The red-tails sat on their eggs hour after hour; the hawkwatchers sat on the hawk bench and watched. Hour after hour they sat and watched the hawks sit. And all for naught. By the middle of May it was over: the nest had failed a second year in a row. The hawkwatchers' high hopes dashed for the second year in a row. Dejection set in.

As ever, Central Park did not fail to work its restorative magic. Apparently every creature *except* the Fifth Avenue red-tails had been exceptionally fecund that year.

All the Missys hatched outsized broods. Bill DeGraphenreid reported that crippled Missy had ten, Missy had twelve, Missy's sister nine. The real egg-laying champion was in a different class entirely: a snapping turtle, who was observed laying between twenty and thirty eggs in a sandy area near the Castle.

Even the wood thrushes had a bumper crop. One year earlier a wood thrush had built a nest in a young cherry tree just off the path to Bow Bridge and produced three healthy fledglings. With the species diminishing in much of the United States, a wood thrush nest in Central Park was an event to celebrate. The year of the second red-tail nest failure, *four* wood thrush babies discovered on June 14 in a nest just west of the Evodia Field were a major gloom-lifter for the bummed-out Fifth Avenue hawkwatchers.

Their joy was short-lived. The very next day the nest was stolen, babies and all. The absence of any nest fragments or feathered remains on the ground below made it clear that the predator had not been a raccoon or rat or blue jay; only a human could have made such a clean sweep of it. Shortly thereafter, as Rebekah Creshkoff was making her birdwatching rounds on her way to work, she taped a hand-lettered sign to the nest tree which declared: "Whoever stole this nest is selfish, thoughtless and cruel!" Taking a more practical tack the next day, she put up a new sign reading: "It is a crime to hurt birds or their nests or to take babies away from their parents!!!! Stealing a wood thrush nest is a violation of the Migratory Bird Treaty Act and is punishable by a $10,000 fine."

Most other birds had better luck: three blue jay babies in a nest with a six-pack holder as its base, located on the path to the Castle; four young downy woodpeckers buzzing all day in a hole near the Gill, demanding food and making it seem as if the tree itself were emitting a mechanical sound. After fledging they were seen at the Azalea Pond all winter.

And more: a single brown thrasher baby emerged from a thorny barberry bush at Cherry Hill—a remarkable feat for a ground-nesting bird in a park with great numbers of dogs wandering around without leashes.

A triumph over nearly insurmountable odds was achieved by a flicker pair that managed to raise three babies in a black locust near the park entrance at 106th Street and Central Park West, thereby confounding the starlings, dread usurpers of flicker nesting holes.

For Central Park birdwatchers, the annual flicker–starling

hostilities are painful to observe. The wily starlings hang around near the flickers' chosen nest site, watching, waiting, biding their time. The large woodpeckers work all day, pecking into a semi-rotten limb—unlike downy or hairy woodpeckers, flickers' bills are not hard enough to excavate harder wood. They drill at the limb hour after hour, tossing out each billful of sawdust with a flick of the head. But as soon as the perfect oval hole leading to the excavated chamber within is completed, just moments before the flickers prepare to settle in—zap! Full frontal attack by the sharp-billed, aggressive starlings. Zip! In flies the starling female, to lay *her* eggs in the hollowed chamber instead of the flickers.

The flickers hardly know what hit them. They flutter around the hole crying and complaining, hour after hour, while the now-resident starling pokes her sharp bill out of the hole and exults. Or so it seems to the infuriated birdwatchers.

A flicker hostile takeover is a horrible event to witness, one that magnifies the human sense of injustice. Sarah Elliott once proposed that the Regulars use slingshots to help flickers win a few of their battles. Of course it's wrong to blame the starling itself. *Sturnus vulgaris,* a European bird, causes no trouble in its native lands, where its ecological niche has evolved over millennia.

The blame goes to the misguided introducer of a non-native species into our ecosystem—in this case, a Fifth Avenue resident named Eugene Schieffelin. He thought it would be a nice idea to have all the birds ever mentioned by Shakespeare available for viewing in the park outside his window. In 1890 and 1891 he shipped in a total of 100 starlings from Europe. Within a half century the starling population of the United States had burgeoned to more than 200 million!

The most thrilling nest of that season was found on May 24th. That was when Charles Kennedy spotted a pair of green herons at the Upper Lobe, the little cove at the northwest corner of the rowboat lake just beyond Bank Rock Bridge. Almost immediately he saw that these birds were not just looking for food. They were bringing nesting material to an oak overhanging the water. A nest! The news created a stir, for these crow-sized, long-necked wading birds had never been known to breed in Central Park.

On June 11, 1840, the green heron (then called the small green bittern) inspired one of Thoreau's more transcendental musings:

> With its patient study by rocks and sandy capes, has it wrested the whole of her secret from Nature yet? It has looked out from its dull eye for so long, standing on one leg, and now what a rich experience it is! What says it of stagnant pools, and reeds, and damp night fogs? It would be worth while to look in the eye which has been open and seeing at such hours and in such solitudes. When I behold that dull yellowish green, I wonder if my own soul is not a bright, invisible green.

In Central Park some 150-odd years later, it was the bird-watchers who spent hours and hours in patient study while the birds were uncharacteristically active, at least in comparison with those in Thoreau's description. In an effort to wrest some secrets from the nesting green herons, Charles Kennedy organized a sunrise-to-sunset vigil—a heron watch.

Unlike the hawkwatch, which was conducted in the heavily populated precincts of the model-boat pond, the heron watch took place in the woods. Unfortunately, the particular woodland that afforded good views of the green heron nest happened to be a nightly gathering place for gay hustlers. When the heron-watchers arrived in the early morning, they had to devote the first few minutes to house cleaning, using long sticks to remove the unpleasant detritus that had accumulated there overnight. At least safe sex was being practiced, Charles said philosophically.

The herons, in the meanwhile, were practicing sex too—of a more productive kind, we hoped. Around 1:30 on May 25th, the heron-watchers were rewarded for their patient vigil with a sight few birders have seen. The female heron was sitting in the nest as Charles Kennedy, Norma Collin, and I arrived for heron-watch duty. Shortly thereafter the male landed on the nest, a green twig in his bill. He sat beside his mate and the two began to sway and nuzzle and intertwine their long, graceful necks. It had the feeling of a stately dance. After a few moments he transferred the twig to her bill; she accepted it ceremoniously and began to weave it into the nest. Then, once again they swayed and nuzzled. Her neck could be seen vibrating and swelling, the feathers erecting somewhat. Her bill was slightly open as he mounted her. Their lovemaking, we agreed, seemed more tender than the hawks slam-bam-thank-you-ma'am-type mating behavior.

One afternoon as I was sitting on a rock across from the heron nest, dreamily watching through my blind of overhanging trees and hoping something unusual would happen, I heard a slight sound that made me look down. There, a few feet from where I sat, a big raccoon slowly emerged from an ancient culvert partially obscured by Japanese knotweed. And right behind her came five small babies, maybe two or three weeks old. One of them wandered toward me and, in that fearless way of very young animals, clambered right over my outstretched feet, leaving muddy paw prints on my white kneesocks.

On May 26th incubation began in the green heron nest. It was a long and anxious four weeks. Then, before any sign of life was apparent in the nest, the herons' changed behavior gave the heron-watchers the clue they'd been waiting for—

the same behavior change we'd vainly hoped for at the hawk nest. Up-and-down movements at the nest's edge, as if feeding some living creature within. At last, on June 23rd, the first two chicks were seen, and the next day three more: five beautiful fluffy green heron chicks in the nest at the Upper Lobe. All five fledged in perfect health and even made the front page of *The New York Times.* We almost forgot our disappointment over the Fifth Avenue hawks.

FALL

Confusing Warblers and Awesome Hawks

❧

When the fall migration began that year of the Fifth Avenue red-tails' second failure, many former hawkwatchers, now Regulars once again, threw themselves with new fervor into the challenge of identifying confusing fall warblers. Why confusing? Because the ranks of avian migrants are increased in the fall by great numbers of recently hatched, dull-plumaged young. These are heading south for the first time, and a large percentage of them will not make it. (Nor would anyone want them to—if all hatchlings were to survive, it would be wall-to-wall birds in the world.)

Most of these first-year birds, as they are called, retain the duller plumage of the nestling phase and generally resemble the female of their species. The males will not acquire the bright-colored feathers their fathers display—breeding plumage—until the following spring. In addition, many of the adult males molt their bright feathers at the end of summer, taking on the duller coloring of the females and juveniles until the following spring.

For the birds, duller plumage is an advantage: it makes them less conspicuous to predators. For the birdwatcher, it's an obstacle: birds become harder to identify. The bay-breasted and the blackpoll warblers, so different from each other in the spring, are almost impossible to tell apart in the fall. The fall bay-breasted warbler has darker legs than the fall blackpoll, says Roger Tory Peterson in his famous *Field Guide*. But not always, he adds. Well, the two look almost identical in the fall, I'm here to tell you, and what's more, they closely resemble the fall version of yet another warbler, the pine. Confusing is hardly the word for fall warblers—maddening is more like it.

The earliest fall migrants arrive in midsummer. That year the Louisiana water thrush was first, as usual. A warbler that feeds on the ground rather than in trees, teetering as it walks like a miniature sandpiper, it was discovered on July 22nd by Norma Collin in its usual spot: the mud flats below Willow Rock. The yellow warblers started arriving on July 26th, a female redstart was sighted on July 29th, as well as a spotted sandpiper and a palm warbler on the same day.

By the first week of August the trickle was turning into a steady flow: black and white warblers, more redstarts, flocks of chimney swifts. The first capital-letter bird to appear in the Bird Register for the season, a BLACK-BILLED CUCKOO, was seen in a willow tree near Bow Bridge on August 5th. On September 7th a capital-letter-with-two-stars-and-heavy-underlining bird was spotted by four birdwatching ornithologists from the American Museum of Natural History. It was a **CONNECTICUT WARBLER**, the most elusive of all elusives, a drab little ground-feeder that likes to skulk in high underbrush. The museum experts often spend their lunch hours in the park directly across the street from their institution, but they don't always share their sightings in the Bird Register. They did this time, drawing an elaborate map of the bird's location. By the end of the day most of the Regulars had found it too.

The fall songbird migration was just beginning to wind down when birds of prey became the center of attention once again. This time it was not the two amorous red-tailed hawks but hawks of all species—hawks and vultures and eagles, thousands and thousands of them.

❧

FALL IS THE TIME for a different kind of hawkwatching than the model-boat pond gang practices—counting and keeping track of great numbers of raptors in the course of their southward migration. Northbound hawks tend to disperse over a wide migratory path in the spring. But on the journey back in the fall hawks concentrate most predictably over certain natural formations, such as mountain ridges. That is where they are likely to find rising columns of hot air known as thermals, which allow them to save energy by soaring rather than arduous flap-flap-flapping.

The hawks flying over Central Park have no natural mountains there to help them on their way. But they have thermals nevertheless, created by the man-made ridges of tall city buildings. That is why fall hawkwatching has always been a part of the park's birdwatching scene. But before Sharon Freedman's hawkwatch no one had ever made a systematic study of diurnal raptors—hawks, eagles, and vultures—flying over Central Park on the way to their winter destinations.

Belvedere Castle, the highest spot in the park, has always attracted birdwatchers looking for hawks. On crisp October or November mornings in the past, little groups would settle down on the steps at the top of Vista Rock, an outcropping of dark gray Manhattan schist at the Castle's base, to look for migrating birds of prey.

The birders would sit on the ancient rock and chat about

whether the new Zeiss 10×40s are superior to the Bausch & Lomb Elite 10×42s, or whether yesterday's smallish Cooper's hawk had actually been a female sharp-shinned hawk. (Since female hawks are significantly bigger than males, the female of a smaller species can be the same size as the male of a larger one.) All the while they'd scan the horizon with their top-of-the-line or their not-so-hot-but-saving-up-for-better binoculars, waiting for someone to call out "Hawk up!" and give a location.

When the cry was heard, everyone would focus on the designated spot and watch a hawk or vulture or eagle—one or many of them—soaring and gliding and occasionally flapping southward, sometimes so high they were beyond the point where they could be identified, and sometimes so low you could make out their field marks without the help of optical equipment.

For at least ten years Sharon Freedman had joined the hawkwatchers at Vista Rock, picking up tips on how to identify hawks in flight. From old-timers like Marty Sohmer, one of Central Park's Big Gun birdwatchers and a particular whiz at hawk-identifying, and from books, she slowly acquired the exacting skills of hawkwatching.

The idea popped into her head in early September of the first year the red-tails nested on Fifth Avenue. She was sitting on the rocks with the usual hawkwatchers when it struck her: Why not make it official? There were plenty of people to help. She herself was unemployed and could put off job-hunting for a few more months. She'd start that very day and continue until mid-December, keeping careful records to submit to the central clearinghouse of all organized hawkwatches, the North American Hawk Migration Association. That was the beginning of City Hawkwatch.

For the Fifth Avenue red-tail addicts, Sharon's hawkwatch

filled in the void left by the departure of the beloved pair at the end of the summer. By the middle of August the Castle replaced the hawk bench as their community's magnetic center.

ORGANIZED HAWKWATCHING IN the United States, and perhaps anywhere, first began in 1934 at Hawk Mountain in Pennsylvania. It was designed to save hawk lives by replacing a "varmint shoot" traditionally held at that site, with observation and methodical record-keeping.

Today there are thousands of hawkwatches at locations along migratory flyways throughout America. Their purpose is partly scientific: to provide data about avian population patterns. In the 1950s hawkwatches provided early warning of diminishing hawk populations owing to DDT spraying, and in recent years counts are indicating the return of several endangered species.

Mainly, however, hawkwatching is a sport. It requires sharp eyes, persistence, and endurance, skills that may be transferred to other parts of life. Like any sport, hawkwatching has its competitive elements, yet it inevitably creates a spirit of camaraderie among its participants.

The major challenge of a hawkwatch resides in identifying just exactly what kind of bird it is way up there. For although a bald eagle may look conspicuously different from an osprey or a turkey vulture when pictured in a field guide, it is a different story when the bird is flying so high that to an untrained eye it resembles a small speck.

Sharon Freedman's skill at identifying hawks in flight is impressive. Using birders' vernacular for two eminently confusable species, the sharp-shinned hawk and the Cooper's hawk, Sharon gives an example of how she operates: "I go by gestalt a lot rather than field marks. You have to at those distances. For instance, with sharpies and Cooper's I'm looking for how they fly. The sharpie is a very delicate flyer and the wingbeat comes more from the wrist. The Cooper's has a heavier wing and the wingbeat comes more from the shoulder. There's a whoomph to it."

Starting at 9:00 or 9:30 in the morning and going until mid- or late afternoon (depending on weather conditions and whether the hawks are flying that day), every day from August 15th to mid-December Sharon and her fellow hawkwatchers look out from the upper observation deck of Belvedere Castle (when it's open) or from the top of Vista Rock, identifying and counting migrating hawks. Of the Regulars, Tom Fiore and Charles Kennedy are Sharon's indispensable assistants—they are there almost every day. Many of the old-timers stop by periodically, including the Big Guns who once gave her identifying lessons. Nick Wagerik is a frequent visitor to the castle. Norma Collin usually drops in for a spell when the day is not bright and glary. (Norma suffers from migraine headaches, and glare seems to bring them on. That is why she almost always wears dark, dark glasses and an eyeshade, even on a cloudy day.) And of course the general public stops in regularly—tourists, school kids, curious passersby—they come and often they come again.

Sharon sits in a high deck chair while her helpers stand nearby or sit in one of the turret-like corner structures that students of medieval castles call bartizans. Surrounded by guidebooks, thermoses, sandwiches, and extra clothing, she announces each arriving hawk to her fellow hawkwatchers.

For purposes of quick identification, she has nicknames for the high-rise buildings visible on the horizon: Tall Red, Pyramid, and the Finger, a sliver skyscraper resembling a rude gesture. Sinai stands for Mount Sinai Hospital's Annenberg Building, a massive structure to the northeast which was properly chastised by the *AIA Guide to New York City* for "thrusting itself onto the skyline like the town bully."

In the first year of her hawkwatch Sharon had a secret goal that sustained her through the sometimes excruciatingly tedious hours of waiting on the Castle deck for hawks to show up: She hoped to see every one of the fifteen species of diurnal (daytime) raptors known to migrate regularly through the northeast quadrant of the United States.

She racked up four on her very first day: kestrel, osprey, broad-winged hawk, and sharp-shinned hawk. She had a turkey vulture on September 6th, a Cooper's hawk on the 17th, a lovely little falcon called a merlin on the 19th. Two goshawks, the largest of the accipiters, checked in on the 28th. Eight species for September.

By the end of October she was up to thirteen species, having added the northern harrier, red-tailed hawk (in migration—she didn't count the resident pair), red-shouldered hawk, peregrine falcon. There was a rare sighting of a rough-legged hawk on October 26th. November 8th was a red-letter day. Two bald eagles flew directly over the park's Great Lawn between 12:15 and 1:00 p.m. "I'm in ecstasy," she was overheard saying as she recorded her sighting in the Bird Register, adding three asterisks before the species name to make sure no one missed it.

By the end of November Sharon's tally was 2,424 migrating birds of prey representing fourteen species. Missing, and a long shot by anyone's calculations, was the golden eagle, *Aquila chrysaetos,* a large dark eagle with a golden crown, less widespread in the East than the bald eagle. Immature birds of the species have brilliant white patches on the wings.

The hawkwatch was due to end on December 15th. On the evening of December 8th, my telephone rang. It was Sharon.

"At first I thought the huge bird coming in from the northeast around Sinai was a bald eagle. Then I saw a black terminal band in the tail, which isn't bald eagle. Then it banked as it came closer and I could really see the white in the wings—the white wing patches . . . of an immature . . . golden . . . eagle."

"I was numb afterwards," Sharon related at the end of her dramatic account. "I didn't know whether to laugh or cry. It was just awesome. Now nothing's missing from the list. Now Central Park has had everything you can possibly get in the northeast quadrant of the United States."*

---

*The following year Sharon added a sixteenth species to the list—a black vulture. According to Roger Tory Peterson's *Field Guide,* this is a southern species whose range ends in southern New Jersey. Its range is clearly expanding northward, for Sharon has spotted at least one black vulture on each subsequent hawkwatch.

FALL TO WINTER

Secrets of a Saw-Whet

End of November. Fall migration over. Fifth Avenue Hawk Show closed for the season. The birdwatching doldrums are at hand. This is the time the little band of Regulars begins to think Owl.

Owls are dream birds. People get worked up at the sight of them. It may have something to do with owls' unusually large eyes and upright posture, giving them an oddly human aspect. Partly it's the opportunity an owl affords of getting a good long look at a bird sitting still instead of one flitting maddeningly from branch to branch.

Above all, it's the joy of the hunt that quickens birdwatchers' hearts during owl season, the pleasures of searching for something hard to find and then, occasionally, finding it.

Central Park's owl hunters look for thick cover. They scrutinize evergreen and holly where an owl might find cover once deciduous trees have lost their leaves. They look for telltale traces, accumulations of whitewash droppings, for instance, that a bird will leave when it has settled in a long-term roost. And pellets, those peculiar by-products of owl digestion that contain so many owl secrets.

Though seven species of owl have been recorded in Central Park since record-keeping began in 1886, only two of these show up with any regularity: the long-eared and the saw-whet owl. A snowy owl was seen in the park only once. That was in mid-December 1890. The short-eared, barred, and eastern screech owls have been sighted two or three times during the last thirty years. The great horned owl is an infrequent visitor.

All these owls are in the family Strigidae. The barn owl, a member of the Tytonidae family, is also rare, though in 1991 one of this weird, monkey-faced species took up residence for almost three months in the Evodia Field, a small meadow just northwest of the Azalea Pond.

Central Park's owl hunters focus on habitats known to entice the two usual owls. Long-eared owls prefer taller pines, and they usually roost near the crown of the tree. Not always, however. Recently a pair of long-ears chose a blue spruce at the base of Cedar Hill for their winter headquarters. They often perched about halfway down from the treetop, though they were still hard to find: their plumage blended perfectly with the tree's mottled trunk.

Saw-whets, happily for birdwatchers, tend to roost lower down, usually in thick evergreen trees or shrubbery, especially ivy or privet or yew or hemlock. This allows the successful owl hunters amazingly close views of the bird, too close indeed for the close-focus range of most binoculars.

The Regulars have been around long enough to know almost every evergreen that has ever harbored a Central Park owl. These trees and shrubs now have their past history reiterated, like Homeric epithets, whenever one goes by. On a winter's day whenever one passes the large hemlock just past the Boathouse, for instance, it is automatically scanned for that telltale dark shape, and someone is sure to announce: That tree sheltered a long-eared owl for a week back in 1988. (The tree won't be there much longer. Like most of the park's hemlocks, it is infested with a tiny aphid-like insect, the woolly adelgid, that sucks the sap from the branches and slowly destroys the tree.)

Just a little farther into the Ramble there are several thick holly bushes and a peculiar tree called a wahoo. Saw-whets have been discovered in each of these and in the two hollies west of the Azalea Pond. If you happen to see someone poking his head into a thick bush, looking up and down and around, you have probably come upon a saw-whet seeker.

It is usually sound that leads to the discovery of an owl in Central Park—not the owl's sounds but the cries and calls of blue jays, crows, or other birds in the vicinity, who are driven by instinct to set up a clamor in the vicinity of a bird of prey. This activity, known as mobbing, can make life miserable for a poor owl who is trying to sleep in a secluded tree somewhere, but it is a godsend for birdwatchers.

Who spotted the owl? Glory is conferred on the sharp-eyed owl-finder. Indeed, the discoverer of an owl is an epic hero, at least briefly, among the park's birdwatchers, and like new comets or diseases, a newly found owl is often referred to by its spotter's name: "Murray's owl" or "Charles' saw-whet."

A few years ago the first saw-whet owl of the season was found on November 29th roosting in the upper reaches of the Shakespeare Garden. This is a hilly plot composed almost entirely of plants mentioned in the works of . . . you guessed it. The smallest of our northern owls, the saw-whet is only about 8 inches long—the size of a robin. It was found by Merrill Higgins, who works for the New York City Department of Correction at Rikers Island and takes pictures of Central Park's wildlife on weekends. (He was to become an impassioned Fifth Avenue hawkwatcher the following year.) While looking around for a bird to photograph, he glanced into a hemlock and couldn't believe his eyes: there was a little saw-whet, at eye level! Merrill's saw-whet roosted in the garden for almost a month.

Why is it called a saw-whet? There are two theories about the name. The more common one claims that the bird's harsh and creaky note sounds like a saw being sharpened or whetted. Audubon himself supports this one. In reality, the bird's song is rather melodious, a whistled note repeated over and over: Toot-toot-toot-toot.

The second explanation was offered by Julio de la Torre, former president of the Linnaean Society and author of the book *Owls, Their Life and Behavior.* He proposes that the word "saw-whet" is a corruption of *chouette,* the French word for a small owl. *Chouette* is the very word used by French-speaking inhabitants of Nova Scotia to refer to the saw-whet owl, a common bird in that province. Since many French Canadians pronounce *chouette* as if it began with an "s"; it ends up sounding like "sou-wet."

As the tiny Shakespeare Garden owl returned to the same roost day after day for most of December, it began to gather a regular audience of human admirers. Members of its own class were less delighted by its presence in their vicinity, and regularly tried to persuade it to leave. On December 14th, for instance, at 7:50 a.m., an angry mob composed of two blue jays, two white-throated sparrows, a titmouse, and a female cardinal flew in to protest the owl's invasion of their turf. They arrived together and proceeded to yell and scream from an adjacent hemlock. They hopped around from branch to branch as they kept up their clamor, occasionally making little mock sorties at the owl. They reminded me of the angry villagers in old horror movies, arriving at the castle with torches and weapons to protest the experiments of the resident evil genius.

During the mobbing the owl opened his eyes and looked at

his tormentors. But he did not move. After a few moments the noisy birds departed, and the little owl went back to sleep, glistening in the morning sunlight.

Lovely as the tiny bird unquestionably was, its admirers knew that the saw-whet owl is not an innocent vegetarian. This is a bird that kills for a living. There are other birds of prey in Central Park: the red-tails, of course, kestrels, peregrine falcons, and an occasional goshawk, Cooper's or sharp-shinned hawk stopping by to sample the local cuisine. But these are diurnal birds whose predatory deeds are performed in full daylight.

Owls work at night, which is why they are usually seen sleeping during the day. Contrary to legend, they can see quite well in the daytime, but they are especially adapted to hunting in the dark. Their retinas have a high concentration of light-gathering cells—the photoreceptors called rods—that contain a special light-sensitive pigment extraordinarily sensitive to the minutest amount of light. Their hearing, too, is highly developed. Even in complete darkness the asymmetrical openings of their ears help them home in on the tiniest sound with precision. They are able to process sound and spatial information three times as quickly as diurnal birds. In addition, structural modifications of the first primary feather on each wing allow owls to fly more silently than other birds. Formidable hunters, owls.

THE FAITHFUL BAND of birdwatchers gathered daily to pay homage to the saw-whet. But they also wondered: What was this little owl eating for dinner in Central Park? Though bird books state that saw-whets were primarily rodent eaters, there was some evidence to the contrary.

Joe Richner was the first to suggest that the saw-whet was an ornithophage. "They don't only eat mice. They eat birds," he declared in mock horror, noting that over the course of the sixty-eight years he has been coming to the park he had seen saw-whets with songbird prey many times. "Chickadees and titmice," he specified.

Nick Wagerik reported seeing a saw-whet sleeping in one of the Ramble hollies the year before with a dead white-throated sparrow in its talons; its breakfast would be at hand—or, rather, at foot—when it woke up that evening.

The Shakespeare Garden saw-whet was obviously eating *something* there. But what? Since owls hunt at night and sometimes in complete darkness, their hunting habits seem safe from human discovery. Yet there's a way to unearth a saw-whet's secrets.

Owls eat their prey without much preparation. Down the hatch goes the dead rodent or bird, all of it—fur or feathers, bones, teeth, tail, and all. The owl's superbly efficient digestive system extracts all the nutrients from the ingested animal and forms the indigestible parts into a compact pellet. Four or five hours after a meal, the pellet is regurgitated whole—for owls the activity is called pellet-casting. Since the gastric juices are somewhat less acidic in owls than in other pellet-casting birds such as hawks, most of the victim's bones remain intact inside an owl pellet.

At the end of a night's hunting, owls often return day after day to the same daytime roost, a sheltered spot where they are relatively safe from predators. There, directly underneath the perch, anyone wishing to discover the secrets of an owl's night activities can usually find a freshly cast pellet containing the skull and bones of its prey.

THAT'S WHAT BROUGHT ME to the American Museum of Natural History one afternoon in December: I had found an inch-long brownish-gray cylindrical pellet in the Shakespeare

Garden, directly under the saw-whet's roost. I hoped it would help identify what the owl had eaten for its dinner that night. A telephone call to the museum's mammalogy department led me to Clare Flemming, a scientific assistant there. She invited me to her lab the next day.

It was the week before Christmas, and Ms. Flemming, an appealing young woman dressed in faded jeans and an outsized white shirt, met me in front of the museum's traditional origami-covered Christmas tree. She has long, wavy, honey-blond hair and remarkably large eyes—though not, of course, for an owl. Hers are light blue, while most owl eyes are yellow or orange. A few species like barn owls have dark brown or black eyes.

Since Ms. Flemming is a mammalogist, and since saw-whets are said to be rodent hunters, she had prepared for my visit by bringing out, for comparison purposes, five specimen boxes with the bones of five rodents that might possibly be found in Central Park. I say "might possibly" because there does not exist any reliable census of Central Park mammals, or, indeed, of most of the living creatures of the park. Only because of the activities of dedicated birdwatchers over the years is there considerable data about the park's bird population. Occasionally a new subject of interest brings in new information about other animal groups. During Nick Wagerik's years of intensive butterfly hunting, he has doubled the number of butterfly species known to be found in Central Park.

After consulting a two-volume reference book called *Walker's Mammals of the World,* fourth edition, Ms. Flemming narrowed down the likely candidates for the former owner of my pellet bones to five: shrew, meadow vole, white-footed mouse, house mouse, and Norway rat, all common rodents in nearby areas. To be sure, there was the unhappy possibility that the bones would not be mammalian at all, but avian. In that case, said Ms. Flemming, I would need to consult someone in the ornithology department.

I was putting my money on the house mouse. A few years ago one had scared the daylights out of me in my kitchen. So I knew that this species, at least, was present in the vicinity. I also knew that Norway rats are present in Central Park. I see them at the Azalea Pond feeding station all the time, generously sharing *their* birdseed with the various sparrows, cardinals, and other birds the seed is intended for. But the rat in Ms. Flemming's specimen box was at least as big as the saw-whet owl. I thought such a rat might eat a saw-whet rather than the other way around.

Using an instrument called a jeweler's pin vise, which is, more or less, a long pin embedded in a holder, Ms. Flemming probed the pellet. Inspecting a bone at the very surface of the still-intact cylinder, she announced immediately that it was not a bird bone. Bird bones are hollow. Here were bones suitable for mammalogical analysis.

The mammalogist proceeded to analyze. She worked slowly and delicately, one might almost say tenderly, moving aside the clotted, brownish-gray fur that formed the surface to get to the miniature bones.

She did not exhibit the expected dispassion of a scientist. As she drew out each bone with a long pair of tweezers, brushing off the fur adhering to it with a paintbrush, she made various exclamations of delight or surprise: "Boy, this is compact!" she'd say, or "Oh, it's like a dream to have real hair on a bone." Ms. Flemming usually works with fossils. There was no question that she was enjoying herself.

As she worked, she watched what she was doing through a stereoscope, a type of microscope that illuminates and enlarges small objects. After scrutinizing each bone and identifying it for me—there were eighty-seven of them—she deposited it into a little white box nearby.

"Here's a metapodial. That's a foot bone. You can see one side is flattish and the other side is ridged. Now, this next one is interesting. It's an ulna—that's the lower-arm bone. See how the end of it is unfused to the bone. That indicates an immature animal." So we knew one thing about our victim. Whatever it was, it was young.

"Oh, here's my favorite bone!" Ms. Flemming exclaimed at one point. The bone she was inspecting is called the astragalus. It forms the ankle, between the tibia and the fibula. I asked why it was her favorite bone, and she answered, "Look," and held it for me under the stereoscope. "Isn't it beautiful?" Hard as I looked, it seemed quite an ordinary little bone to me.

We went through tiny ribs, many phalanges (the fingers and toes, as she explained), and quite a few tail vertebrae. "We can see this animal had a long tail," she noted. "None of these bones are diagnostic," she added, the scientists' way of saying, "Don't know yet."

"Oh, here's the pelvis," she said with delight, and showed me a bone that resembled one of those monocles with a long handle that Eustace Tilley carries in *The New Yorker* magazine. She brought over the box with shrew bones and showed me that animal's pelvic bone. The shape of the opening was completely different. "Now we can eliminate the shrew."

"Oooh, I think we have a sacrum here! That's the three fused vertebrae forming the hip," Ms. Flemming explained. "I think we can now eliminate the meadow vole." She inspected the vole box and beamed, "Yes, the meadow vole sacrum has four vertebrae."

We had eliminated the rat from the very start. The bones in the pellet were obviously far smaller than any of the bones in the rat box. Consequently, we were down to two possibilities: the house mouse, *Mus musculus,* and the white-footed mouse, *Peromyscus leucopus.*

I was still betting on the house mouse, for nobody had ever recorded white-footed mice in Central Park. John Hecklaw's "Central Park Wildlife Inventory," a pamphlet published by the Central Park Natural Resources Group in 1984, does not list *Peromyscus leucopus* among the mammals of the park.

Another hour passed as Ms. Flemming extracted bone after bone from the pellet. When she drew out a mandible and compared it with both the house mouse and white-footed mouse mandibles, the odds began to change. "See the way the angle of that lower incisor is sharp on our specimen?" she said. "It is also sharp on the *Peromyscus,* but not quite as sharp on the *Mus* mandible. This is beginning to look more like a *Peromyscus.* Still . . ." She did not seem quite convinced. "No teeth," she noted as she put the mandible into the little white box. "Too bad. Teeth would be diagnostic."

Ms. Flemming had been poking and probing for almost three hours. Finally, she drew out the very thing that would provide us with a definitive diagnosis. The skull. "This should do it," she said.

The little skull was a miracle of miniature perfection. "Now this is *really* beautiful," I said, hoping I wasn't hurting Ms. Flemming's feelings about the loveliness of the astragalus. As she pointed out the various features of the mouse skull, she brought over a human skull for comparison. I shuddered—it was the first real human skull I had ever seen up close.

"Here are the bullae," she said, and showed me two tiny cavities in the specimen skull. "They house the inner ear." The bullae of the human skull were amazingly similar, though obviously larger. "And here's an inner-ear bone," she exclaimed, returning to the pellet's skull. She pointed out a tiny little bone —the stapes. She turned the skull over and showed me two vertical slits just south of where the creature's large incisor was still attached to the end of its jaw.

"See those two slits?" she asked. "Those are the inter-incisive foramina." Foramina, she explained, are openings. "Look at the shape of those slits on the *Mus* skull," she continued. "Very different, aren't they? Now here's the skull of the white-footed mouse. See the two slits? What do you think?"

It was perfectly clear: Our mystery bones came from a white-footed mouse, a creature heretofore unacknowledged as a resident of Central Park. Now we knew it was there. I felt like stout Cortez upon a peak in Darien.

Ms. Flemming affixed a label to the little box containing the collection of pellet bones: "*Peromyscus leucopus,* White-footed Mouse." Then she gave it to me. I offered to donate the bones to the museum's collection, but she declined. I put the box in my pocket quickly, just in case she changed her mind.

As we walked from her office to the elevator, we passed many cabinets filled with amazing bones of amazing animals: anteaters, armadillos, and an extinct carnivorous marsupial called a thylacine. None of them, however, seemed as desirable as the bones that had once been inside a saw-whet owl roosting for almost a month in a hemlock tree in the Shakespeare Garden in Central Park, bones that were now in a little white box in my pocket to keep forever.

WINTER

The Christmas Count

❧

The second winter of the Fifth Avenue red-tails was one of the coldest in decades. But on December 18th, the day of the annual Central Park Christmas Bird Count, winter took a brief break; it was a mild, almost springlike day. Most of the Regulars were at the appointed spot—the south pump house of the Reservoir—by the appointed hour, 8:00 a.m.: Norma, Charles, Tom, Mary Birchard, Mo and Sylvia, Harold Perloff. Missing was Nick—he and the butterflies are not morning creatures. Also Sharon Freedman, who was celebrating the end of her successful hawkwatch with a visit to friends in Bucks County. Altogether, about fifty people showed up.

A Christmas Count is an annual bird census that takes place throughout the United States during the final weeks of December. It was started in 1900 as an alternative to traditional "side hunts," during which hunters went out after Christmas dinner, chose sides, and competed to see which group could shoot more birds. Today's more merciful event is sponsored by the National Audubon Society and undertaken by volunteer bird-lovers who count not only the different species of birds but also the actual number of individuals. This can provide useful scientific data about bird populations. It is also a fund-raising event for the Audubon Society.

There is a competitive element as well: each group going out to cover an assigned section of the park hopes to come up with something special—a rare or unusual bird—with which to amaze the gathered birdwatchers when the results are tallied at the end.

Sarah Elliott was in charge of the Count that year—her ninth at the helm. "I have a red-headed yellow-capped Sarah," someone announced as she approached at 8:00 a.m., a bright yellow cap jauntily perched on her bright red hair.

The hardest part of the Count is the assignment of territories. Every section of the park must be covered, with an experienced birder in charge of each, but some areas are far more desirable than others. As usual that year, everyone wanted to do the Ramble or the North Woods, the two best bird habitats.

By far the least promising territory in the entire park is the southwest quadrant. It is the only section that does not include a single body of water. It has no woodland area, just little clumps of trees here and there between heavily used stretches. The bulk of it is taken up by Sheep Meadow, a vast expanse of twenty-two acres originally intended as a parade ground for military exercises. But military activities were banned on the premises even before the park opened. The sheep were installed instead. (The sheep were exiled to Prospect Park in 1939.)

Sheep Meadow during a Christmas Count reminds me of a word game I used to play with my kids when they were young: Fortunately-Unfortunately. You make up two sentences, the first beginning with "Fortunately," the second, which must be related to the first, beginning with "Unfortunately." For example: Fortunately, a good Samaritan offered the starving man a hamburger. Unfortunately, the starving man was a vegetarian. Well, fortunately, Sheep Meadow always contains hundreds of birds—but unfortunately, all of them are totally uninteresting—gulls and pigeons. I had done the southwest quadrant on the Christmas Count two years earlier and can't say it was a thrilling experience.

Nevertheless, Richard Rabkin, an officer of the Linnaean Society and a kindly fellow, volunteered to be in charge. When no one else offered to join him, Tom Fiore came to the rescue and signed on. I'd go birding in the Sahara with Tom, and so I became the third member of that team. A brand-new birdwatcher named Mary filled in a fourth slot. She was worried about her lack of birding skills, but we snagged her with the idea that she could be useful counting rock doves. She laughed when we explained that rock dove is the official name of a bird she knew well, the pigeon, and she joined our team.

We certainly had pigeons—450 was our final total. We also had 150 house sparrows, a conservative estimate—nobody bothered to count too diligently. And we had large numbers of a bird that birdwatchers would never, *never* call a seagull—there's no such bird as a seagull, they'll insist. Our gulls were ring-billed gulls, 225 of them, all standing around on Sheep Meadow. By 9:35 a.m. the southwest quadrant was living up to its lowly reputation.

As we passed a stand of white pines near Pine Bank Arch, one of Central Park's thirty historic bridges, we heard a welcome clamor: a bunch of blue jays screaming "Thief! Thief!" The cries were accompanied by the angry chirps of a little collection of house sparrows. A moment later six blue jays and four sparrows converged on one of the pines, making a major ruckus.

We hurried over to investigate, for a noisy avian mob usually means one thing—a bird of prey in the vicinity. I hoped it would be one of the Fifth Avenue red-tails—no one had seen either of the pair for a few days, and a Christmas Count sighting would be a good omen. Tom, however, seemed more excited than a red-tail sighting warranted. He had a sense that this would be a "good bird," perhaps because thick evergreens are not what our hawks usually chose to perch on.

It proved to be a good bird indeed, the best bird of the count. Roosting deep in one of the pines was a long-eared owl, a much-desired winter visitor to Central Park. In the daytime, roosting owls are usually asleep. This bird was wide awake, and it peered out of its evergreen hideout directly at us.

It was poetic justice. With this long-eared owl we knew we humble southwest quadranters were going to be the champs. We jumped and hugged each other and then, since crowds of birdwatchers would soon be trooping over for a look at the owl, we searched for a good viewing spot to recommend, one that would offer a reasonable view of the owl without unduly disturbing its peace. We found a perfect viewing window on the east side of the cast-iron bridge itself, and we marked the exact place for viewers to stand with a pair of twigs.

As we were taking our last lingering looks at the owl, a well-dressed family arrived at the bridge, tourists taking an early morning stroll through Central Park. How did we know they were tourists? They were carrying maps and guidebooks and wearing cameras. Besides, like many out-of-towners in Central Park, they had a special look of curiosity and amazement mixed with a soupçon of fear. They stopped and openly stared at our little group as we leaned over the side of the bridge and peered with binoculars into a tree.

They were disbelieving when we offered an answer to their

politely unspoken but all too obvious question: What in the world are you looking at? "We're looking at a long-eared owl," we volunteered.

They were the Carwells from Bowling Green, Kentucky: Sally and her children Amy and Charlie. They had come to see the Big Apple and its Christmas sights: the animated store windows at Saks Fifth Avenue, the tree at Rockefeller Center, the Christmas show at Radio City Music Hall—and the Central Park Zoo, which is where they were heading. The idea that one of the sights New York had to offer was a long-eared owl roosting in a white pine had never entered their minds.

Sally and Amy accepted our offer and looked through binoculars at the owl. It took them a while to focus the glasses on the right place, but eventually they succeeded. "Oh my goodness gracious!" exclaimed Mrs. Carwell. "Oh wow!" said Amy. Charlie passed up his chance at a look—he was a cool pre-teenager. Sally Carwell then said the words we so often thought to ourselves, words that made this encounter only slightly less gratifying than our encounter with the long-eared owl itself, and the subsequent applause of our fellow bird-watchers when we announced our find at the tally. "A wild creature right in the heart of New York City! Isn't that remarkable!"

WINTER TO SPRING

Roostwatch

❧

The stars were bright in the city sky as I left my house for Central Park. Yet none of the winter constellations were out, though it was February, not the brilliant-eyed Charioteer nor Orion the Hunter with his three-studded belt. And where was Sirius the Dog Star, brightest of all heavenly objects in the winter sky? The earth had spent the long night rotating on its axis and now, as I set off for Central Park at the unearthly hour of 5:00 a.m., the stars of a different season had risen in the winter sky: the giant star Arcturus as bright as a little moon overhead, and Altair, Vega, and Deneb, the great triangle of the summer sky, shining almost as radiantly as on an evening in July.

Not only the stars seem out of season when one is out in the city before dawn. There is a languid, almost tropical feel to the place at that hour, like Bogart's Key Largo before the hurricane strikes. You know that the storm is coming, that in an hour or two those hushed empty streets will be full of cars and trucks and taxis and buses with maddened drivers blasting their horns, trading insults about each other's mothers and fathers; you know that those empty sidewalks will be full of people bumping into each other while failing to apologize. All hell is about to break loose, but at five in the morning the storm seems far away.

The streets were almost deserted as I walked toward my destination that morning: only a newspaper delivery man removing bundles from a car trunk, a doorman smoking outside an apartment house, and an old man sweeping in front of an all-night grocery to give a hint of the workaday world. Here and there another passerby, someone coming home from somewhere—a night of illicit love, an all-night vigil at a deathbed; someone on the way to somewhere—a train to catch, an early shift as short-order cook at a Greek coffee shop. I caught myself staring at one of those dawn walkers, wondering why he was out at such an hour, what his story was. I laughed to think what he'd say if he knew mine.

IT BEGAN A FEW WEEKS EARLIER, on a freezing Sunday afternoon in January as Norma Collin was making her daily rounds of the Ramble. That's when she heard a loud tattoo coming from a nearby tree.

"I know that sound," she thought cheerfully, and within minutes she had found her bird: a downy woodpecker, the smaller and more common of Central Park's two resident woodpeckers (the red-bellied is the other, though a few hairy woodpeckers and flickers sometimes overwinter). The little bird was in a half-dead black cherry tree, drilling away at a short, stubby snag about ten feet from the ground.

The woodpecker was familiar, but its action was puzzling. It was not just drumming as woodpeckers often do, sending a message of love or war to another of its kind, nor was it extracting larvae or grubs from crevices in the bark. It was clearly making a hole.

Clinging to the reddish-brown, scaly bark with feet especially adapted for feeding on vertical surfaces—two toes pointing forward and two back—zygodactyl is the scientific term—while bracing itself with its stiffened, spinelike tail feathers, the woodpecker attacked the branch with a steady, rapid back-and-forth motion. It looked for all the world like a living black-and-white jackhammer. The cavity was already quite deep, and every so often the bird would briefly disappear within for a second or two, and then reappear to toss out billful after billful of sawdust, with a jaunty flick of its head at each toss.

What in the world is this bird up to? Norma wondered. It wasn't odd to see a woodpecker making a hole exactly the way this one was doing. But that was to be expected at the start of the breeding season, in April or May. Why was this bird excavating a hole in January? As she watched, she noted that the woodpecker was definitely a *he:* on the back of his head was a bright red patch, the only feature that distinguishes male downies from females. Well, maybe he's practicing for spring, she finally decided, and left it at that.

Norma had been observing the woodpecker for a good fifteen minutes and her fingers and toes were numb. She was looking forward to getting home and settling down with a nice cup of linden tea—she had picked the blossoms herself from a tree on Pilgrim Hill the previous spring. First, however, a quick stop at the Boathouse to write down her day's sightings in the Bird Register. During the excitement of the migration seasons a bird like a downy woodpecker is hardly even mentioned in the long lists of visitors that fill the Register's pages. But January is a month when the park's residents—woodpeckers, sparrows, blue jays, crows, titmice, and cardinals—take center stage. The downy woodpecker drilling a hole near lamppost 7631 was by far the most important sighting Norma had to report that day.

The next day Charles Kennedy, voracious reader of bird

books and accumulator of arcane information about almost any natural history subject, came up with a possible explanation. He happened to have, among his huge number of bird books, one called *Woodpeckers of Eastern North America* written by an engaging amateur naturalist named Lawrence Kilham. That's where Charles got the idea that Norma's woodpecker must have been making a winter roost hole.

Roost holes, according to Kilham, offer a woodpecker insulation against cold and shelter from the wind. When the outside temperature is 17 degrees Fahrenheit it can be 11 degrees higher inside a roost hole, even higher if the woodpecker's body fits the cavity exactly. "The amount of energy . . . conserved may make the difference between survival and death during periods of extreme weather during winter," Kilham wrote.

A woodpecker's winter roost hole has advantages for birdwatchers as well: it offers anyone wishing to study the Picidae family an opportunity to locate its members dependably. "Just wait until an hour after sunrise, and the chances of locating a Downy or a Hairy in a reasonable time can be slim. . . . But in taking the trouble to be by a roost hole at dawn, I have had some of the best of birdwatching," wrote Kilham.

Yes! thought Charles, lover of eccentric projects. If Norma's discovery turns out to be a roost hole, we've got some serious woodpecker-studying to do.

IT WAS ALMOST 5:30 A.M. on that February morning by the time I arrived at Naturalists Gate, the park entrance at West 77th Street. The other five birdwatchers were already waiting and we entered the park at once. Though it was almost an hour before sunrise, day was beginning to break, and we didn't want to be late. Crossing the Upper Lobe at Bank Rock Bridge, we took a short path going south, and then east, taking us under the Rustic Arch with its buttresses of natural stone. Then we were in the woods, in the wilds of the Ramble.

A few moments later we arrived at the scraggly cherry tree. There we put down our knapsacks, took out our thermoses of coffee, and settled in for a wait. Roostwatch had officially begun.

The woods were silent and the roost hole was barely visible in the semi-darkness. Good. We'd arrived in time—every bird in the park was still asleep. Now we would keep a watch to see if a downy woodpecker was roosting within.

We waited and watched the day grow brighter. Around 6:00 a.m. we heard the first bird sounds—blue jays screaming in the distance. The roost hole was silent. Sunrise arrived at 6:24 a.m.—no sign of life at the cherry tree snag.

In truth, few of us had faith that there was a bird inside that neat hole, not even Charles, who is a dyed-in-the-wool optimist. I wasn't convinced that there even *was* such a thing as a roost hole, in spite of Kilham. After all, there had been downy woodpeckers in Central Park for as long as anyone could remember. Wouldn't someone have seen a roost hole by this time—Tom Fiore, for instance? And yet we must have had more than a little room for hope. Why else would we have been willing to spend all that time standing and waiting, our toes and fingertips freezing and puffs of smoke coming out of our mouths with every breath as the temperature hovered around 20 degrees?

Our little band waited on. If you get up at 4:15 in the morning you don't give up just like that. At 6:40 a.m. Charles said rather quietly: "Well, look who's there." The rest of us were deep in conversation. We looked up and saw a soft, fuzzy head filling up the roost hole entrance. The bird was decidedly morning-sleepy.

First the woodpecker stared straight ahead without moving, as if in a trance. After a minute or two he began to look up and down and up and down, more alert. Finally he seemed to fix his gaze directly at us. "What? You guys?" I imagined him thinking. "I have to put up with the sight of you and your binoculars all day long in the park. Now you're here when I get up in the morning too!"

He flew out—whoosh! It happened so fast we couldn't entirely believe what we'd seen. Was it a group delusion? No, everyone finally agreed, we had really seen a bird zip out of the hole. Charles looked at his watch. It was exactly 6:42 a.m.

The woodpecker alighted on a nearby dead branch and commenced a loud tattoo, a long unbroken drumroll. There were ten bursts of drumming, which Charles diligently timed. Each lasted about six seconds. Experts can identify a woodpecker by its drum pattern. The hairy woodpecker's roll is shorter and louder than a downy's, with a greater interval between each stroke. The yellow-bellied sapsucker starts with a short roll and ends with five or six distinct taps. Sapsuckers, by the way, appear in Central Park only during migration. Their grid-like rows of neat holes (for collecting sap) may be seen on the trunks of many park trees.

An active roost hole! Usually the excitement of birdwatching is based on unpredictability, for unpredictability breeds hope. As you stroll around with binoculars at the ready, you never know when something new and exciting, perhaps something rare or beautiful might show up. That's hope.

Now the pleasure was in the very predictability of the bird. Predictability breeds hope too, we discovered, the same sort of hope that each year's cycle of seasons inspires. It is somehow deeply fulfilling and hopeful to know that the phoebe will arrive in Central Park on March 13th every year, give or take a few days. Or that if you stand at a certain place at a certain time a particular bird will show up and perform a predictable action—like zipping in or out of a hole in a tree.

A small group of Regulars began to monitor the roost hole with regularity, both at dawn and at sunset when the bird flew in for the night. The essence of Roostwatch was timekeeping, making note of the bird's precise moment of entry and exit. This produced a fact, a concrete piece of data. Charles Kennedy, the major instigator of the project, loved facts. Above all he loved to time. He timed how long cardinals bathed in his brook under Balcony Bridge. He timed owl fly-outs, and mockingbird songs. He also loved to measure—the sticks in the ill-fated hawk nest, for example.

On March 16th, with an orange-red full moon setting in the west just before sunrise, woodpecker wake-up was at 6:01 a.m. On the evening of March 27th and the following morning, March 28th, the roostwatchers managed to document a woodpecker's full night of sleep. They watched the bird scoot in at 5:55 p.m. They returned the next morning—it hardly seemed worth going home, somebody quipped—and observed downy reveille: it was at 5:46 a.m., one minute before sunrise. The bird had been in the roost hole for eleven hours and fifty-one minutes, Charles Kennedy announced.

On April 9th, a cold spring day, a new development at the early morning Roostwatch warmed the watchers' hearts if not their toes. A moment or two after the woodpecker emerged from his roost hole at 6:38 a.m. (the bird was not sleeping in that morning—Daylight Savings Time had commenced), another downy woodpecker materialized as if out of nowhere. Double vision. Only one bird had flown out of the hole, and now there were two woodpeckers climbing on a nearby branch. They looked identical but for one detail: no red patch on the newcomer's head. It was a female. "Yahoo!" exclaimed Charles, the eternal romantic.

After that, the female showed up regularly at Roostwatch, morning and evening. Before bedtime the two birds were often seen feeding in the vicinity, busily working the bark of nearby trunks and branches for whatever it is that woodpeckers extract from crevices—seeds, bugs, larvae. Then, as night began to fall the female would fly off somewhere—to her own roost hole, no doubt—and he would zip into his. In the morning the female was often on a nearby tree, waiting for her sweetheart to rise and shine.

April 28th was a warm, balmy spring day. Six roostwatchers had assembled for the evening roostwatch by 6:30 p.m. This time they did not reminisce about past vigils at the roost hole, those cold February and March waits. Tonight the talk was about the present, about the day's birds.

Spring migration was in full swing and the day had been a glorious one: scarlet tanagers, rose-breasted grosbeaks, an orchard oriole, and a yellow-billed cuckoo had been sighted in the Ramble. Tom Fiore had heard a singing indigo bunting at 6:20 a.m. that very morning. The electric blue bird had been near the Belvedere Castle at the same place indigo buntings show up every year—near a patch of Kentucky bluegrass favored by that species. Tom had also sighted a least flycatcher making his "che-bek che-bek" call, the only way you can distinguish this bird from four other small, virtually identical flycatchers. Nineteen species of warblers had been seen so far that spring, and Tom had seen sixteen that very day.

Little leaves were already out on the scraggly cherry; soon they would obscure the roost hole. Somebody picked one and crushed it to demonstrate its bitter smell—cyanide! That evening the woodpecker took the longest time arriving.

He didn't show up until 7:30 p.m., later than ever before. The female arrived almost immediately after. The pair first flew from branch to branch of a nearby tree, making little chip sounds and feeding desultorily. At about 7:35 the female suddenly disappeared, and at 7:43—whoop!—into the hole whizzed the male. This time there was a round of applause from the assembled gang.

Night had fallen and the Regulars walked out of the park together. No matter how many times it had happened before, it still seemed incredible, incredible to have penetrated the secret life of a creature so *other*.

On the morning of May 3rd four watchers waited from 5:30 a.m. to 7:00 a.m. We waited and waited until the day was bright and clear, as clear as our knowledge that no bird was going to come out of the hole. Quite a few other birds showed up in the vicinity that morning—a great crested flycatcher and a wood thrush singing its enchanting song. Also seven species of warblers. Everything was lush and green. The knotweed was high as an elephant's eye. But the roost hole was empty.

The downy woodpecker and his mate failed to show up at the roost hole that evening, and the next. Gone. Obviously they had other business to take care of. This had been a winter roost hole, after all, and the winter was over.

ACT III

Hawk Mania

The sight of any free animal going
about its business undisturbed, seeking its food,
or looking after its young, or mixing in the company
of its kind, all the time being exactly what it ought to be
and can be—what a strange pleasure it gives us.

SCHOPENHAUER

SCENE ONE

What's Past Is Prologue

Pamela Clauss lives on the fifteenth floor of Woody's building, two floors below his penthouse. Though she is not a birdwatcher, one spring she found herself watching birds regularly, two huge birds that had begun to perch almost every day on the ledge outside her living-room window. They were obviously hawks. She knew they had a nest nearby, for she'd watched the pair flying back and forth past her window with sticks in their beaks. But she did not know that a crowd of hawkwatchers were gathering in the park every day to follow their every move.

She didn't know how extraordinary it was for red-tailed hawks to build a nest on a city building, Indeed, she didn't even know these were red-tailed hawks. But she knew that these were exciting birds and looked forward to their visits.

When she read my article about "her" hawks in the *Wall Street Journal,* she looked me up in the phone book and gave me a call. It was clear that she could see more intimate details of hawk behavior from her window than the hawk-bench gang would ever see, even with their most powerful binoculars. For every morning she had breakfast with one or another of the hawks—she at her kitchen table with her coffee and toast, the hawk on the window ledge hardly a foot away, also enjoying its morning meal. As Ms. Clauss described it:

"They'd usually arrive with a bird, a dead bird, of course, which they would tear to pieces. It was interesting how they ate: first they'd pluck all the feathers off—drop them off the ledge onto Fifth Avenue. They'd usually eat the head whole, but they would never, never touch the entrails or the claws—they'd leave those on the ledge for me to clean up.

"They did leave a mess, but I never minded. They brought mice and rats and all sorts of things to Mr. Allen's penthouse too—we chat about the hawks when I meet him in the elevator. He says he likes them. He doesn't seem to mind the mess either.

"I got so fond of both of the birds, the huge, big one—he was here much more often—and the smaller pale one, I think she's the woman." Here I interrupted to tell her about reverse sexual dimorphism among hawks. Then she continued:

"They'd look at me sometimes with those wonderful big eyes, sort of 'What are you doing here?' and I'd look back as if to say 'What are *you* doing here?' Sometimes they'd make little chippy noises at me, and I'd make chippy noises back at them. I loved those birds. They were gorgeous. I was so happy to be able to watch them."

Ms. Clauss was far from happy on July 19th, when she saw workmen on a scaffold approaching the nest. She was frantic.

"The minute I saw the scaffolding going up, I thought: 'Oh no! They're going to disturb the nest!' " she told me in another telephone call. "I went across the street to talk to the super, but the only person there was Johnny, the handyman. And he told me there were only two families in residence in the whole building. Everybody else was in Europe or off in other vacation spots. Even the super was away. So nobody would know what was happening.

"I didn't know what to do. So I called the Fish and Wildlife Service to see if they could save the nest. I talked to someone named Mr. Weber there. And now, suddenly, the nest is gone. I'm afraid I didn't do any good at all."

But Ms. Clauss was wrong about her call. Eight months later it produced results.

❧

JUST AS THE HAWKS were beginning their second attempt to raise a family at the Fifth Avenue nest site, Special Agent Kevin R. Garlick of the U.S. Fish and Wildlife Service, Division of Law Enforcement, sat down at his desk in Lawrence, New York, to write a severe letter. It was to Alvin Traub, the managing agent of the hawk building.

> This correspondence will serve as written notification that the U.S. Fish and Wildlife Service has identified the nest of a pair of red-tail hawks on a building managed by your organization. . . .

Alvin Traub could not have been happy to read these words. A hawk nest on his building at 74th Street and Fifth Avenue the previous year had attracted a lot of attention. The building's board of directors did not like attention. In addition, he had been beleaguered by calls from birdwatchers in various states of excitement (I was one of them), and several communications from building residents expressing concern one way or another. Even Mary Tyler Moore had called. It was just another thing to worry about. When the nest was removed during the building's major façade cleaning that June,

I imagine he had hoped it would be the end of the whole messy business.

Not that Mr. Traub was unmoved by Nature. He loved animals, especially dogs, and had once had a Yorkshire terrier he had been very attached to. But dealing with his Fifth Avenue clientele was hard enough without hawks causing problems.

Now it looked like the birds were nesting again. Alvin Traub read on.

> It was brought to the attention of the U.S. Fish and Wildlife Service that [last year] building maintenance removed a red-tail nest from the building. By doing so, a violation of the Migratory Bird Treaty Act was committed.
>
> Raptors (hawks, falcons, owls, and eagles) are protected by the Migratory Bird Treaty Act, 16 USC section 703, which states: "Unless and except as permitted by regulations, it shall be unlawful at any time, by any means or in any manner, to pursue, hunt, *take,* or capture, kill, *attempt to take* . . . any migratory bird, any part, *nest,* or egg of any such bird, or any product, whether or not manufactured, which consists or is composed in whole or part, of any such bird or any part, *nest,* or egg thereof. . . ."
>
> This is a strict liability statute that requires no degree of culpability or knowledge for a criminal conviction that can result in a fine up to $10,000 per bird, nest, egg, or part thereof.

Those last words said it loud and clear. Though the building was obviously not short on funds, this could add up! If they fined him for one nest and three eggs, that could mean $40,000 right there. And what did the words "part thereof" imply? Was every damn feather in that nest worth another ten grand?

The letter went on solemnly:

> The service has a stewardship obligation to the American people to ensure the continued survival of all raptors.... Please refrain from removing migratory bird nests in the future....

The communication had an imperious tone Alvin Traub was not unfamiliar with from his years of dealing with Gold Coast dwellers. He called Special Agent Garlick first thing the next morning. The conversation brought immediate relief. There would be no fine this time, Garlick announced, as a monarch showing mercy to a subject might do. But please be informed that the U.S. Fish and Wildlife Service was very aware that a pair of red-tailed hawks had built another nest this year at the exact location. The service would not be lenient again if this new nest came to any harm.

"You can be absolutely certain we will not touch that nest," the managing agent assured Mr. Garlick.

All this is to explain why the Fifth Avenue red-tailed hawks would not have to start constructing their nest from scratch when nest-building time rolled around the following year. Thanks to Special Officer Kevin Garlick (and thanks to Pamela Clauss, who alerted the Fish and Wildlife Service to the violation), the previous year's nest remained in place. For the first time, the birds would be able to build their nest on a foundation of old twigs. For the first time, a thicker layer of nesting material might cover the sharp anti-pigeon spikes.

If the hawks returned to their Fifth Avenue site for a third try—a big if—they might have a better chance at success.

❧

OCTOBER 14—3:45 P.M.—Answering machine message from Norma Collin (sounds excited):

Amazing news! I was walking near Bank Rock Bridge this morning with Annabella, Harold, and Mary when we heard a big commotion. One minute later we saw a red-tailed hawk pursued by five or six screaming blue jays. It was clutching the biggest, fattest rat you ever did see. The hawk was flying from tree to tree trying to find a good eating roost. Finally, it landed on a pretty low branch of a red oak and waited until the jays decided to leave her in peace. We definitely thought it was a female, because she was so large.

Well, she put on quite a show. First she pulled off some fur, then she attacked the belly. Mary and Annabella just squealed "Eeeeoooo!" And Harold kept a running commentary. "Look, there's the heart, there's the liver," and so on.

And then we saw the band. A bird band on her right ankle! We'd never come across a banded bird in the park before, at least I hadn't. We tried terribly hard to read it, but we couldn't make out a single letter or number. Couldn't even see if there *were* any numbers on the band, except I guess there have to be. Finally, we had to leave—we'd been there almost an hour—but the hawk was still there when we left.

A BANDED HAWK in Central Park! Where was Chocolate? No one had seen her since July. And where was Pale Male? Sightings of the light-colored hawk had been few and far between since the failed nesting attempt.

A period of confusion set in for the next two months. One of the park rangers believed that the banded bird was a male who had ousted Pale Male in a savage territorial battle. He claimed to have seen Pale Male in late October with a broken primary on his right wing and an injury to his right eye. But Norma believed that the banded hawk was a female and so did Anne Shanahan. Both of them continued to report occasional sightings of an uninjured Pale Male.

Two months later, on Christmas Day as it happened, Anne

Shanahan had a close encounter with the banded hawk, a decisive one. For this time the bird was not alone. "A light-colored red-tail and a larger, darker one were eating a pigeon just west of children's playground at 77th Street," she wrote that day in the Bird Register. "The larger, darker hawk had a silvery band on right leg."

Anne knew the light-colored red-tail as well as she knew her little dog Bijou. It was Pale Male. The larger hawk was obviously his mate, for hawks do not tolerate the presence of strangers in their territory, especially not while eating.

Once again there was talk about First Love. Six months earlier, just as the pair were beginning to sit on eggs at the nest, Tom Fiore had had a close-up view of the female's ankles. No silvery band visible on either, he was sure of that. Now Pale Male's mate was sporting a conspicuous band on her right ankle. "This one *has* to be First Love," Charles Kennedy declared.

Charles and a few other Regulars—I was one—loved this idea—we loved the circularity of it—back to the beginning. But besides the romance, it seemed so logical that this was the first mate. We knew she had been banded. Now we thought: How can there be *two* banded females with a connection to Central Park? Though two years had passed since Len Soucy had banded and released First Love over the Great Swamp, it could have taken her that long to regain her health and recapture her predatory skills. Then she would have flown right back to Central Park to reclaim her mate.

Norma Collin thought it was a new female. There was a reddish orange glow to the breast that neither Chocolate nor First Love had had, she pointed out. Others agreed and began to call the banded bird Butterscotch.

Only Anne Shanahan resisted these ideas. She had a powerful feeling that in spite of the silvery band on her ankle, and in spite of the reddish glow, this hawk was the same bird as last year's female. Birds, she knew, can change their overall look after their annual molt. (Like most birds, red-tails renew their feathers each year, a process usually completed well before the end of summer.)

Whether this was Chocolate, the old mate, or Butterscotch, the new, or whether this bird might possibly turn out to be First Love back with Pale Male at last, the answer to the mystery resided in the silvery band on the female hawk's ankle.

SCENE TWO

A Tourist in Trouble

Pale Male and his mate with the silvery band spent all January making public appearances at their old haunts: the Octagonal Tower, the Green Shade Building, Woody's—a good sign that they were planning to nest. The hawkwatchers began to show up at the hawk bench again.

On Sunday, February 5th, the day after a snow- and sleet-storm rain left the park covered with a thin coating of treacherous ice, two occasional park birdwatchers, Tony and Netti Luscombe, had a bird sighting that temporarily diverted attention from the Fifth Avenue hawks.

The Luscombes had just returned to New York from a long stay in Peru where Tony had run a conservation agency for ten years. Eager to resume Central Park birdwatching, they'd decided to check out the Pond that morning, whatever the weather. The Pond is a little body of water at 59th Street and Fifth Avenue, just across from the Plaza Hotel. It occasionally attracts interesting water birds—that year a pair of wood ducks had been hanging around there most of the winter.

It was bitterly cold. The Luscombes found the Pond almost completely frozen, except for a patch of open water under Gapstow Bridge, the stone arch that spans its north end. The wood ducks were there, together with some thirty mallards eating bread a good Samaritan was tossing down from the bridge.

Suddenly Tony's attention was arrested by a bird on the ice in the middle of the pond. "What the devil is this?" he said to Netti. They both focused their binoculars on a gull. It was sitting and nibbling on a crust of bread.

There were a couple of crows, a single grackle, and a lot of other gulls on the ice, all ring-billed gulls, the common winter gull of Central Park. But the bird Tony and Netti were now staring at with growing disbelief was not a ring-billed gull.

Tony wiped the lenses of his binoculars and looked again to be sure. But there was no mistaking this gull. Uniformly slate gray, with a faint ring around its eye and a black bill, it was clearly an adult gray gull, *Larus modestus.* Netti agreed.

The Luscombes had seen gray gulls before. Indeed, they had seen thousands and thousands of these birds running in and out of the surf as they snapped up mole crabs, their favorite delicacy. But that was at Villa Beach, just outside of Lima, Peru. Peru is the northern boundary of the gray gull's range. Central Park is 3,676 miles farther north than that.

"What the devil is this bird doing here?" Mr. Luscombe asked Netti. She had no good answer. Though birds sometimes get blown off track by storms, this seemed beyond the realm of possibility.

When the Luscombes returned home, they looked through various reference books to make sure there wasn't a local gull that might remotely resemble the bird they had seen on the Pond that day. Perhaps it was a young bird of some North American species, since immature gulls are notoriously difficult to identify. But none of the birds pictured in the books, neither young nor old, remotely resembled the one on the ice that day.

It was time to activate the birders' grapevine. Tony called

me and I called Sharon Freedman, who was just listening to the evening news.

She interrupted in the middle of my description of Tony's bird. "Wait a minute! I just heard that a huge aviary at the Bronx Zoo collapsed this morning and a whole bunch of rare South American birds escaped," she said. "Tony's bird must be one of them!"

The next day's papers gave the details: At 10:45 on the morning of February 5th, the zoo's arch-shaped, wire-covered outdoor aviary had collapsed under the weight of heavy snow accumulations. Constructed in 1899, the aviary was one of the institution's first exhibits. Its interior re-created a rugged coastal island where aquatic birds could swim, nest, and fly freely, and it was home to a colony of 100 South American seabirds—Magellanic penguins, Inca terns, and guanay cormorants, as well as Andean, band-tail, and gray gulls.

When the aviary collapsed, thirty-three of these rare seabirds took off into the wild blue yonder, dispersed by winds that gusted up to fifty miles an hour. Most of them had never known any other home but the Bronx aviary, where they led a carefree life, with regular meals and tender care dispensed by the zoo's ornithology department. Among the missing birds were twelve Inca terns, one band-tail gull, twelve Andean gulls, and eight gray gulls.

Though it is a seabird, the gray gull nests far from the ocean, indeed, in one of the world's most inhospitable locations: the Atacama Desert of Chile. Inhospitable is putting it mildly: It rains there on an average of once every seven years. During the nesting season gray gulls must commute sixty or more miles to the sea each day, for food and water. Now one of these creatures was sitting on the ice at the Pond in Central Park.

Tony had reported a bluish-green band on the bird's right leg, which helped Dr. Donald Bruning, the curator of birds at the Bronx Zoo, identify the individual bird. It was a male, one of five gray gulls the zoo had imported from Peru in 1988.

The thought of a hand-fed, pampered zoo bird out in the cold, hard world of New York City made me uneasy. While human visitors often exaggerate the dangers that lurk in this city, avian tourists have genuine reason to fear: The resident birds are urban toughs. Ring-billed gulls are aggressive contenders for available food; crows are ferocious fighters; and there are other notable predators around, the resident red-tailed hawks in particular.

On my way to the Pond on Monday morning, I walked by a huge white trailer containing dressing rooms and rest rooms for the crew of a movie being shot in the park that week. The "honey wagon," as it's known in trade vernacular, was parked in the carriage road just east of the Pond, and its driver, George Gowing, was standing outside it as I went by. He was from Kansas City, Missouri, and this was his first trip ever to New York City.

"New York is nothing like what I was told it was going to be like," Mr. Gowing was eager to tell me. "Everybody is so nice and friendly." I hoped the gray gull would find an equally warm welcome here, though I had my doubts.

When I reached the Pond, Tony had already arrived. The gull was still there! It was hunkered down in the middle of the ice, looking hungry and miserable, I thought. I saw it get up and nibble at a crust of brownish bread someone had tossed out earlier; at once a crow waddled over and snatched the tidbit away.

Tony waited while I went to call some birders. When I returned, he headed for the Central Park Zoo nearby, and within minutes he was back with Anna Marie Lyles, its associate curator of animals. She took one look at the bird and said, "Yup, that's him." Then she went back to the zoo to plan the capture.

At a little before 11:00 a.m. Ms. Lyles returned with three zookeepers. One of them had a net. Another had a metal pail full of little fish. He had just finished feeding the penguins and was going to lure the gull toward shore with penguin food. But as he tossed the shiny fry toward the gray gull, nearby ring-bills quickly grabbed each one. All this time the rare bird from the Bronx just sat on the ice and watched. After a while Ms. Lyle and the keepers departed. They would try again later, when the gray gull might be hungrier.

Tom Fiore arrived around noon. He had never seen a gray gull before except in the zoo. He knew he couldn't add this bird to his life list, that important tally of bird species serious birders like to keep: zoo escapees don't count. In any case, enlarging his life list was not the reason Tom had come. He simply wanted to see a gray gull. He looked at the bird for a long time, incorporating into his mental ornithological data base the bird's field marks, shape, size—everything that would help him identify it at another encounter. Then he and Tony went off to check the Lake and the Reservoir for other escapees.

I was alone with the gray gull at 12:45 when I heard a loud bird commotion: blue jays screaming and crows cawing. I was not the only one who understood what this alarm meant: A nearby flock of pigeons dispersed in a frantic cloud and sparrows scattered for shelter into nearby bushes. All the gulls on the Pond took off as well, the aviary fugitive among them.

Moments later, as I expected, a red-tailed hawk sailed into view—unmistakably Pale Male. He circled the pond and then landed on a tall tree in the Hallet Nature Sanctuary at the west end of the Pond. He looked hungry—his craw was flat. For once I was not delighted to set eyes on him.

It was freezing cold. By two that afternoon the hawk was gone, and some of the gulls had returned. But the gray gull was not among them. Norma Collin and Harold Perloff came by at around 3:00 in search of the rare visitor. Anna Marie Lyles and the three keepers came back at 3:45, and then again at 9:30 the next morning. But the bird had vanished. Perhaps it was heading for Peru, or perhaps it was just another New York tourist who had come to a bad end. We'll never know.

SCENE THREE

The End of Doubt

Wouldn't you think the hawkwatchers might have simmered down and become philosophical about the Fifth Avenue hawks after two years of disappointment? Not at all. As the annual drama resumed, we worried more than ever.

We worried about crows. We worried about last year's rat poison as well as this year's, for we'd read about a class of chemicals called xenobiotics—it includes herbicides, insecticides, and all the other -cides—that may have long-term effects on birds. If ingested, these substances cause decreased fertility and diminished hatching success while seeming to leave the bird unaffected.

We worried that after two failed attempts Pale Male and his mate would abandon the Fifth Avenue nest and choose some more conventional nest site. We'd never have such an entry into their secret lives if they nested high in the crown of a leafy tree. The hawk bench, in the meanwhile, offered an unimpeded view of their aerie at 74th and Fifth. We could follow their every move once again if they returned to that site. If if if!

In mid-February I returned from a trip to find a long message from Norma on my answering machine:

Great news! Today at about 3:30, when I came into sight of the Fifth Avenue nest, the female hawk was *on the nest!!!* At 3:35 Pale Male flew in. Then they both flew off the nest and onto the antenna of the Octagonal Building. They mated on the railing there—about six seconds, and then he flew off and she sat on the antenna for about fifteen minutes or so. Then he came back and they mated again, for about six seconds again. So here we go again!!!

Another thing that was really funny. Pale Male went over to Woody's building. Just the floor below Woody's penthouse there's an ornamental tree on someone's terrace—I think it's called a weeping birch. He flew and landed right on top of it and I thought: Boy, that's a silly perch. And don't you know he clipped a branch from that tree and carried it over to the nest! Those people are probably nursing this treasured tree. So that was absolutely marvelous. We were all giggling over that.

Then, around four o'clock, they were both on the nest. She sat there with her back toward him, on the south side, and he was on the north side. Somebody, I don't remember who, said, "Maybe they're not getting along."

Six seconds! Wasn't that a world record? All the hawk-watchers reconvening at the hawk bench agreed there were special reasons to be hopeful this time. But that hopefulness, as usual, was tempered by anxiety. *This* time the hawk apartment people would get out their brooms and clean off the ornament above their middle window.

We decided to get in touch with Mary Tyler Moore again. Her presence as an ally had calmed our nerves in the past. Why not send her a note to let her know the hawks were back and needed her protection?

Dear Mary:

We thought you might like to know that the red-tailed hawks are once again nesting on your building. And though they did not succeed in raising a family on their first try or their second, we have quite an optimistic feeling about this latest attempt. Here's why:

The nest, as you know, is built on the sharp anti-pigeon spikes that protect the ornament above the middle window on the 12th floor. Both the first year, and the second (because the building removed the nest after the first nesting attempt) the hawks had to build directly on top of the spikes.

Subsequently the Fish & Wildlife Service warned the Managing Agent that it was a violation of the Migratory Bird Treaty Act to remove the nest. Consequently, the nest was *not* removed after the second nesting attempt.

And so, because they did not have to start from scratch this year, we believe that this time the nest will have a firmer foundation, which should improve the chances of successful incubation. (It could be that the sharp spikes damaged the eggs during incubation. Now they'll be covered over better.)

We hope we're not going into more details than you might want to hear, but we know you are a nature lover, so we're hoping you still have a lively interest in this story.

All best wishes!

The Central Park Birdwatchers

❧

ONE SUNNY MORNING in the last week of February a small group sat at the hawk bench, exploring the perplexing identity question: Was the bird with the silvery band a new female,

Butterscotch? Or could she be Chocolate, as Anne Shanahan secretly believed? Or would she turn out to be First Love?

Nick Wagerik had joined the hawkwatchers that morning while waiting for the day to warm up—the first mourning cloak butterfly of the year had been known to appear in late February and he didn't want to miss it. He listened to the discussion for a while, and then politely but firmly objected to our practice of assigning names to the hawks—anthropomorphism, he called it, an attitude he held in disrespect. Since Nick is held in the highest respect by the birdwatching community, nobody took serious issue with him that day.

But Nick (I should have said to him), surely there is a case to be made for anthropomorphism, that maligned yet unavoidable practice of attributing human emotions to other orders of animals—unavoidable, at least, for those who spend time closely observing wildlife. It's not that animals have emotions like ours. It's that *our* emotions resemble those of other animals. For doesn't evolutionary evidence show that all human characteristics with survival value have precedents in the phylogenetic past? Clearly such supremely valuable human properties as reasoning ability and emotional complexity did not spring forth fully evolved, like Athena from Zeus's brow. The notion that only humans think and feel is a relic of creationism harking back to the Victorian era.

In her book *The Hidden Life of Dogs,* Elizabeth Marshall Thomas went further, suggesting that animals engage in a form of anthropomorphism of their own. When a dog snarls and protects a slimy, dirt-encrusted bone from its human owner, the canine is assuming that the human will find the revolting item as desirable as another dog might. Cyanomorphism, she called it ("cyano" deriving from the Greek word for dog), a dog's equivalent of anthropomorphism.

Another form of reverse anthropomorphism might be called ornithomorphism. This occurs when people assign birdlike characteristics to *Homo sapiens.* Ornithomorphism is seen in various of our turns of phrase, as when we talk about children "leaving the nest," whereupon the parents become "empty nesters." But some of our most common examples of ornithomorphism are dead wrong. "Eats like a bird" to denote a small appetite, for instance. Though a hummingbird may consume only an ounce of combined insects and nectar in a day, this actually constitutes more than twice the bird's total weight—the equivalent of a 150-pound man taking in 300 pounds of food daily.

"I wonder if the hawks have names for us the way we have names for them," Anne Shanahan said at the end of the discussion. She reported that on her way to the hawk bench she had seen the female red-tail perched on the northeast corner of the Metropolitan Museum. A few minutes later she saw Pale Male arrive with a dead rat in his talons. He landed with it directly beside his mate and she immediately reached over and grabbed it. It was a gift, a love gift. He sat beside her quietly while she ate the rat—it took about fifteen minutes. Then, when she finished, the light-colored hawk hopped on her back and they mated. After that they sat there quietly one next to the other for a good ten minutes.

"They were having a cigarette," said Marcia Lowe, a new hawkwatcher that year, as she lit up a cigarette of her own.

❧

MARCH 8TH WAS an unusually mild day—the temperature went up to 63 degrees. Perhaps this sent a signal that spring was on its way, for the first spring songs of the fox sparrow and the junco were heard the next day. Unfortunately the temperature quickly went down to a wintry 27 degrees.

The next day had another first: the first Woody sighting of the year. Nobody meant to spy on him, but in the course of hawkwatching, binoculars inevitably passed by his huge penthouse windows. If Woody happened to be out on his terrace or right next to a window at that exact moment, well, let's say that nobody refused to look. That day he was seen inside his upper window, practicing the clarinet.

Around March 17th the first exchanges were observed: one bird in, other bird out. Incubation had begun. But because the birds had built this year's nest on top of last year's (thanks to Kevin Garlick and the Migratory Bird Treaty Act), the nest was now bigger and thicker; the sitting bird was harder to detect. That bird was mainly seen at exchange time, when the vague lump in the midst of the sticks suddenly sat up and became a red-tailed hawk.

Keeping track of exchanges, the hawkwatchers observed that the female put in about three times as many hours on the nest as the male, which was still considerably more male participation than indicated in the various expert accounts of the species. "Both sexes incubate, although the female, fed by her mate, usually does most or all," write Brown and Amadon. Pale Male was obviously a thoroughly modern male, the hawkwatchers observed. He'd probably change diapers too.

Around 2:00 p.m. on Saturday, April 22nd, my telephone rang. It was Norma, a little breathless: Something was going on at the nest. I biked right over, almost colliding with a taxi on the way.

When I arrived at the hawk bench and pointed my binoculars at the nest, I immediately noticed more activity. The female kept getting up from the incubation position and standing on the edge of the nest. She seemed to be looking down at something. Pale Male soon appeared at the nest with a chunk of something in his talons. He deposited it deep inside and left, whereupon the female stood at the edge, facing in and made up-and-down motions, as if feeding. All of this looked sufficiently promising for Norma to go out on a limb, as it were, and write "HAPPY BIRTHDAY!" in the Bird Register.

At the nest, more of the same for the next few days: food-bringing, edge-standing, up-and-down moving. But no actual sign of new life within. Among the hawkwatchers, much doubt and nay-saying: There's nothing in that nest; it's another false alarm like last year and the year before.

By the third week of April the spring migration was gaining momentum. Warblers were streaming in. On April 22nd, Dorothy Poole saw the fifteenth warbler species of the year. One of the best birdwatchers among the Regulars, Dorothy made her rounds of the park in the early, early morning on her way to work. Dorothy's warbler was the black-and-orange Blackburnian warbler, a "good bird." The next day a cerulean and a blue-winged warbler were first sighted, and an orchard oriole just east of the Castle. More good birds. Nick Wagerik found a black swallowtail butterfly that day, almost three weeks earlier than he had ever seen one in Central Park.

On April 24th, almost a week before the migration was likely to peak, Tom Fiore had a Big Day. His list of sightings took up one and a half pages of closely printed writing in the Register, with a grand total of ninety species of birds. Including the red-tailed hawk, of course, he added at the end.

On Wednesday, April 26th, a birdwatcher named Deborah McMillan brought her Swift 22-power telescope to the model-boat pond. She took a long look at the nest and then called out. Everybody lined up to have a look, birdwatchers and passersby alike.

There were shouts and even tears of joy at what the telescope revealed: two fluffy white chicks opening their beaks and gobbling down tidbits delivered every few minutes by their wildly successful and proud new parents: Pale Male and his permanently renamed partner. Who cared if she was Butterscotch or Chocolate or Banana Fudge Strawberry Ripple. She was now and forever . . . Mom.

SCENE FOUR

The Epic Catalog

Who was there to see the baby hawks on April 26th, that day of days? Who was there that week? What was the scene?

The Regulars were there at the model-boat pond, proud and jubilant. Norma stopped an Argentine couple wheeling their baby in a fancy pram. "We have a baby too! We have two babies!" she exclaimed. Charles whooped. Sharon Freedman and Tom Fiore, Annabella, the Girards—all present, all cheering. Dorothy Poole arrived in the late afternoon—she had left work an hour early. Anne Shanahan was there. She quietly announced it was Audubon's birthday, without revealing that it was also her own. Deborah Allen, bird photographer and poet. Alan Messer, bird artist. Nick Wagerik, after checking the butterflies at the Conservatory garden and the wildflower meadow in the north end of the park—he was there. David Monk, tree lover and polyglot, was there the next day. Mary Birchard was not there. It was a Wednesday and she would not get there until Friday, her day off.

OTHER BIRDWATCHERS CAME. Sarah Elliott, Mo and Sylvia, Max and Nellie Larsen, Marty Sohmer. They stopped at the hawk bench and hailed the babies. Peter Post, a Big Gun, came by, but not because he was celebrating hawks. Why admire birds that kill? he asked. He came for the company. Other Big Guns stayed in the Ramble, fattening their various lists (Life lists, Year lists, State lists, Park lists) with spring migrants. Starr Saphir brought twenty-five birdwatching students to see the babies the following Monday.

Frederic Lilien, a young Belgian filmmaker, was there. He was videotaping the excited hawkwatchers for his documentary about Central Park. By the end of the season he would become a hawkwatcher too and the story of the Fifth Avenue hawks would take over his film.

A new crop of hawkwatchers were there that day: Marcia Lowe arrived first. She had become a big-time admirer of the red-tails that year and a major-league worrier about their health and happiness. Jim Lewis was an advertising consultant—he came between appointments. Merrill Higgins, discoverer of the saw-whet owl and now one of the most passionate hawkwatchers, had to come from work at Riker's Island. He arrived a little before dark. Holly Holden and Blanche Williamson came after work; they were photo editors for *People* magazine.

Holly Holden had been taking a walk in Central Park a month earlier when someone (she didn't know it was Merrill then) asked her an odd question: "Do you want to look through the scope?" She almost said no, she still recalls with amazement. "My initial reaction was: Oh my God, someone is jumping off a building!" But she looked and was hooked.

THE MODEL-BOAT POND habitués were there. Oscar Maxera, a handsome, white-haired artist from Argentina—one of his paintings is at the Whitney Museum—was beaming: "I'm feeling the emotions of a father," he said at his first sight of a baby hawk in the nest; Lorenzo, a minister for Jehovah's Witnesses, dressed in black leather with a sheriff's badge on the left breast pocket, a cowboy hat on his head, he was there in his mechanized wheelchair, his computerized Bible on his lap. He keyed in "birds of prey" and found citations in Revelation, in Isaiah 18:6, and in Genesis 15:11—"And when the fowls came down upon the carcasses, Abram drove them away." Anne Schwertley appeared, an elegant, enigmatic woman. "I ran around with the abstract expressionists in my misspent youth," she murmured.

Edward sauntered by, disheveled seller of *Street News,* with more than a few missing teeth and an observant eye: "Every morning I see the white hawk sitting on the tree across from the Boathouse when I go in for my coffee," he said. He claimed there were actually three hawk babies in the nest, but nobody knew he was right until the afternoon of April 28th. While filming the nest for CBS News late that afternoon, a cameraman named Buddy Tyler looked closely at his video monitor and then called a nearby hawkwatcher over. "How many chicks did you say there were up there?" he asked. The screen revealed a newly hatched chick in the nest. Now there were three.

THE OWNERS OF MODEL BOATS took time off from their expensive hobby (the boats can cost thousands of dollars) to take a brief look at the hawks. Among them, the master of the *Knar,* a replica of an ancient Danish sailing vessel he had constructed himself. Everything on the *Knar* was authentic, he said, including a miniature barrel of grog for the crew. They wouldn't work without it, he explained.

But where was the fat, sulky boy of twelve named LeRoy

and the *Lillian B. Womrath,* his racing sloop? Once he had lost a close race on this very pond against a schooner with a tiny cannon on her foredeck, steered by a well-dressed mouse named Stuart Little.

DOG WALKERS STOPPED BY, attracted by the clamor. Among them were the owners of Benedict, a basenji; Stella, a bulldog; and Woofy, a mutt. Woofy's owner grew up around the corner. She used to walk in the Ramble with her governess back in the 1920s. Also there, escorted by owners, were Maxie, a cairn terrier, and Lucy and Tiffany, west highland terriers.

What was the clamor? Every time Pale Male arrived with a tidbit, every time Mom arose from brooding to fix a twig or feed a chick, the joyful hawk-benchers would call out, "Way to go, Mom!" or "What a dad!" and other such cheers. For the first two weeks Mom continued to brood the chicks some of the time, to sit on top of them protectively, especially when it rained.

Barbara Lazear Ascher arrived with her black poodle Gabriel. A writer, a friendly, beautiful woman, she had a tender feeling about the red-tailed hawks and often hung out at the hawk bench with the Regulars. Her eyes filled with tears when she heard the news. Her grown-up daughter Rebecca arrived ten minutes later with her own black poodle, Rogier. The dogs rejoiced too—not at the hawks' hatching but in each other's company. They romped.

Others of note turned up: Hollywood producers John Starke, John Penotti, and David Sameth, scouting the area for their forthcoming movie *I'm Not Rappaport.* Looking dubious, Sameth squinted into Deborah McMillan's scope and then said, "My God, there really *are* hawks up there!"

• • •

WHAT WAS BLOOMING in the park to welcome the young hawks into the world? The Virginia bluebells were in flower. The lesser celandine and violets were blossoming all over. The wild geranium, windflower, dame's rocket were opening in various gardens and flower beds throughout the park. Too late for daffodils. They were just about gone, a few dingy, withered relics remaining to hint at their former splendor. The garlic mustard growing wild in the woodlands was blooming—its young leaves make a tasty nibble.

The cornelian cherry, or *Cornus mas,* the first shrub to flower in the park except for the witch hazel that blooms in midwinter, had finished blooming: the yellow, flowerlike parts now on the plants were fruits. The Norway maples were in flower, bright yellow-green blossoms many casual passers-by think are fresh spring leaves—a celebratory sprig of these flowers appeared in the hawk nest that very day. (The less sentimental noted that hawks bring flowering sprigs to their nests for parasite control.) The Japanese cherries were almost finished flowering throughout the park; soon it would be snowing cherry petals. Azaleas were just coming out. The crab apples, shadbush, and the redbud below Belvedere Castle were flowering—the redbud's young blossoms are delicious in a salad, says Charles Kennedy. On Balcony Bridge golden lichens were in bloom on the bridge walls—a sign of improved air quality in the park, for lichens will not grow if the air is polluted. Good for baby hawks, fine, clean air!

WHAT MIGRATORY BIRDS were seen that day? Only eight species of warblers on the 26th. "Fairly quiet," wrote Tom in the Bird Register. There had been sixteen kinds of warbler on the 24th.

What birds were beginning to nest? Blue jays, song sparrows, cardinals were defending their territories. Downy woodpeckers were drilling nest holes. Tom saw a pair of green herons at the Upper Lobe near their old nest site. Two pairs of red-bellied woodpeckers were excavating nest holes, and at each work site a starling sat waiting nearby. A pair of mute swans were building a nest on the island in the middle of the rowboat lake. One of the Missys had had eight new ducklings at the Lower Lobe, a quiet backwater officially called Wagner's Cove.

And of course American robins, perhaps the most prolific breeders of Central Park. They were building nests everywhere. Some were already sitting on eggs. Why *American* robin? To distinguish it from the just-plain-robin, a much smaller European bird, the one Blake had in mind when he wrote, "A Robin Redbreast in a Cage / Puts all Heaven in a Rage."

Some of the migrants that nest in the park—the Baltimore orioles, catbirds, red-eyed vireos—were beginning to trickle in. The red-winged blackbirds had arrived in February and were singing away in various reedy patches of the Lake: in the tall, feathery reeds called phragmites at both ends of Turtle Pond, the ones at the Lower Lobe, and those near Bow Bridge. Also a few in the thick stand below Willow Rock. Meanwhile the park's horticulture staff made plans to eradicate all those phragmites, for it is an invasive weed that prevents more bird-friendly plants like cattails from taking hold. All horticulturalists hate phragmites—but red-winged blackbirds, mallards, and other park wildlife make good use of it, perhaps for lack of anything better. A kingbird pair was hovering around a future nesting tree at the Harlem Meer.

What birds took baths that day? A worm-eating warbler near the Azalea Pond. Fifty house sparrows, two blue jays, and a flicker were seen splashing in Charles' brook. Why

Charles' brook? Two years earlier Charles Kennedy had appointed himself keeper of an 88-pace stream (as he measured it) that flows under Balcony Bridge. He devoted hours to clearing it of trash, and undertook to count the bathing birds. A haiku he wrote there:

this evening
the brook finally learned
my reflection

What birds were heard singing, welcoming the baby hawks to the world? The white-throated sparrow—Old Sam Pea-bo-dy; American robins, singing everywhere; a few chimney swifts just in from the Amazon basin, twittering above the rowboat lake; ruby- and golden-crowned kinglets; titmice; American crows (yes, there's a crow in Europe too); red-bellied woodpeckers; blue jays crying "Thief! Thief"; flickers crooning "wick-a, wick-a"—many flickers: it was the height of flicker migration; a mockingbird singing loudly and conspicuously at Strawberry Fields, though the crowds paying homage to John Lennon there didn't seem to hear him. Also heard that day, at 6:00 a.m., a bullfrog doing his version of singing at the edge of the Lake, near Bow Bridge: Jug-a-rum!

What butterflies were seen on the great day? A Juvenal's duskywing arrived to honor the hawks—its earliest sighting in Central Park. Also there that day: five mourning cloaks, an American lady, a painted lady, an unidentified (lepidopteran) lady, two red admirals, one orange sulphur, and eighteen cabbage whites, all seen by Nick Wagerik.

WHAT WERE THE sun and moon doing, what planets were out on the great day the baby hawks were seen? The sun rose at 6:03 that morning and set at 7:45 in the evening. The waning moon, three days from New, rose at 4:19 a.m. and set at 5:01 p.m. Jupiter rose at 10:51 p.m. and would shine brightly until dawn the next morning. Venus, meanwhile, would rise about an hour before sunrise—barely visible at dawn.

The weather was sunny and pleasant all day, a cool 49 degrees for the hawk babies' breakfast but a perfectly spring-like 68 degrees by the time they were tucked in under their mother's downy breast feathers for the night.

WHOM DID I TELEPHONE with the news? Special Agent Kevin Garlick of the U.S. Fish and Wildlife Service was the first. "It wouldn't have happened without you," I told him. I had cowered under his stern disapprobation in a phone conversation the year before, when I admitted that I had handled the nest. "You violated the Migratory Bird Treaty too," he informed me sternly, and it actually took some pleading to avoid further consequences. Now he sounded much less forbidding. "I'll write another letter to the managing agent and make sure they leave the nest up there again after the hawks finish this year," he said. And he did. And they did.

Harold Perloff in England was next—he had left that year at the end of March. The nest was rebuilt by then, but its success was far from clear. "Have we had a blessed event?" were his words when he heard my voice on the phone. "Now I can sleep easy. I haven't slept for a month."

I called Charles Preston, the Denver expert. "A year ago you wrote, 'Better luck next year.' Well, we just had the better luck," I told him.

I didn't forget to call Neil Calvanese, head of horticulture for Central Park. He makes decisions about rat poison. "We'll make an effort to minimize rat baiting while the babies are in the nest," he had said, and in fact rat poisoning had been suspended in all major hawk areas throughout the park during the incubation period. He promised to continue the ban while the babies were in the nest and kept his word.

Finally, I called Len Soucy, hawk humanitarian, at his rehabilitation center by the Great Swamp. "Our hawks just had babies. What do we do if one falls out of the nest?" I asked. It was our very first day of hawk parenthood, and we had already begun to worry.

SCENE FIVE

Fear of Flying

❧

Each new stage of hawk-child development called for celebration. On April 29th one baby stood and flapped its stubby white wings. Hurray! On May 5th a chick toddled to the edge of the nest and sent a projectile stream of white excrement onto the building's green canopy below. Wild cheers from the hawk bench.

The hawk pair revealed an unexpected talent: superb parenting. All day, every day, they went back and forth from the park to the nest, returning with an assortment of prey in their talons, mainly pigeons and rats, with an occasional songbird to relieve the tedium. Rarely did they come home empty-footed.

On the nest they kept a watchful eye out for aerial predators, and on occasion gave chase to crows or other hawks that had wandered into their closely guarded territory. They seemed to have it down to a science: taking turns at the nest, standing guard on nearby buildings, watching, waiting, hunting, feeding. Back and forth, back and forth.

By May 12th the nestlings were beginning to peck at the food themselves, instead of waiting for Mom to feed them little chunks. By May 17th they were beginning to develop their flight feathers. A buffy orange color appeared on their upper breasts, and each was beginning to display a row of dark markings below the chest, a baby belly-band. One of the three was distinctly bigger than the others, and one of the smaller ones had a brighter reddish glow to its upper chest.

As the hawklets grew out of the baby stage, the parents completely stopped feeding them beak-to-beak. Now the chicks could be seen tearing into the food as soon as a parent arrived with prey. Yet they never squabbled. They waited their turns like nice children while one or another of their siblings tore in. The hawkwatchers were surprised by such exemplary manners on the part of young predators.

AS THE CHICKS GREW BIGGER and more active in the nest, telescopes began to proliferate at the model-boat pond. On June 1st there was a Celestron, a Bushnell, a Kowa, and two Nikons all aimed at the nest, with legions of passersby lining up for a look. "*Who* are you looking at?" was a common question, usually accompanied by a knowing leer, prompting Merrill Higgins, owner of one of the two Nikons, to post a sign on his tripod: WE ARE NOT VOYEURS. WE ARE WATCHING RED-TAILED HAWKS IN A NEST.

For foreign tourists—there are more of them in Central Park than New Yorkers—he included:

English: Hawk nest with chicks.
French: Faucon nid avec enfants.
(Later he emended it to *buse à queue rousse.*)
German: Falke Nest mit kinder.
Spanish: Halcón nido con niños.
Italian: Falco nido con bambini.
Portuguese: Falcão ninho com criancas.
Japanese: Taka kadomo.

Merrill's greatest pleasure, his passion, indeed, was sharing his telescope with the public at large. "Want to look in the telescope? Want to see the hawks?" he regularly asked anybody walking by the hawk bench. Amazingly enough, many people declined the offer. Once, out of curiosity, I followed one of the decliners and asked why he didn't want to look. The man answered: "I didn't want to see anything embarrassing."

Merrill's sign headed a few questions off at the pass, but there were always others. Certain ones were recurrent: How did you find the nest? What do the hawks eat? People were especially curious about the people in the hawk apartment. Who were they? Did they know there was a hawk nest outside their window?

We tried our hardest to get in touch with them. Addressing our letters and packages "To the Residents of the Twelfth-Floor Apartment" at the building's address, we sent them a series of communications: photos of the nest taken by various birdwatchers; information about red-tailed hawks from various bird books; a reproduction of a painting of a red-tailed hawk we bought in the natural history museum gift shop; several letters begging for an interview, all promising to maintain their anonymity, if they so desired. There was no response.

At the hawk bench it didn't take long for a newcomer to get savvy. Steve the Teamster was one of the best examples of the instant-expert syndrome. He was a stagehand on the *I'm Not Rappaport* crew now shooting at the other end of the model-boat pond. "The female is the one on the left," I overheard him saying to a newcomer on his second or third visit to the hawk bench. "You can tell because she's bigger. The father's smaller, his color's much lighter, see how pale he is? And see the way the babies are bulging on top? That shows they've just been fed." (He endeared himself to the hawk-watchers by regularly bringing over trays of leftover fruit, cheese, crackers, and doughnuts from the film crew's catering wagon.)

One day we kept track of the reactions at the telescope. The most popular for first-time viewers was "Oh wow!" heard thirty-six times. It was followed by "This is incredible!" (31), "I don't believe this" (8), "Unbelievable" (4), and *"Incroyable"* (a tourist from Marseille). Two people said, "It's like *Nova*." The actress Glenn Close was the third to announce, "It's a miracle."

A telescope was taken down to its lowest height for a five-year-old would-be hawkwatcher, Joanna Seirup. (Her father John is in charge of the Ramble Project, a volunteer effort to pick up litter in the Ramble.) Joanna still had to stand on tiptoe to see the hawks. She did not say a word as she gazed through the scope, but the sight touched her imagination and when she got home she wrote a book. It was entitled *HOK NEST,* and it read from back to front. Using orthography that spelling reformer George Bernard Shaw would have admired, she wrote: "THEY EAT PIJINS. THEY HAV FUZZY HEDS. THEY ARE RED TELD HOKS." Under one of her well-observed illustrations of a bird in a nest she wrote: "THIS BABY HOK IS ABOWT TO FLY."

THE BABY HAWKS were about to fly. The three well-fed hawklets could be seen practicing for flight, flapping their wings and jumping up and down in the nest. They were as big as their parents by the end of May and looked ready to go.

For the red-tailed hawks, incubation begins with the first egg laid. Thus the eggs hatch asynchronously and by the same token the young leave the nest over a period of days. According to Ralph S. Palmer's *Handbook of North American Birds: The Diurnal Raptors,* the young of red-tailed hawks fledge from 42 to 46 days after hatching. Therefore, all calculations pointed to the first week of June for fledger #1 to take off.

Strictly speaking, fledging refers to the development of a nestling's flight feathers (from the Middle English *flegge,* meaning feathered). The verb is thus as unique to birds as feathers are. Since the first flight inevitably follows a bird's complete feather development, fledging and first flight are virtually synonymous.

Being ready for first flight, however, doesn't mean you'll be a flying ace right from the start. The hawk babies had been stuck in an increasingly cramped, buggy space for almost a month and a half, with no opportunities to test their wings. They possessed instinctive skills for flying, to be sure, but their lack of experience could lead to trouble. Among all birds, the great majority of nest mortalities occur just after they fledge.

According to Palmer, a red-tailed hawk's first flight is not much of a journey. Usually it's nothing more than a big hop from the nest to a nearby branch. The young spend the next phase of development as "branchers," typically perching on branches a bit lower than the nest, where they clamor to be fed. They return to the nest for the night. This may go on for up to two weeks, after which they begin taking longer sorties from higher perches. On their branch the young will generally exercise their wings, but mostly, according to Palmer, they persistently beg for food. These young birds are not deemed to be "capable of sustained, confident flights" until nine weeks after fledging.

How would the Fifth Avenue hawklets manage their first flight without the welcoming branches of a tree to catch them? Before they had developed any skills, how would they navigate the perilous journey from their twelfth-floor nest, across Fifth Avenue and all its traffic, and into the safety of Central Park's trees? A child passing by the hawk bench looked up at the nest and asked, "Can hawks swim? What if they fall in the boat pond?" The hawkwatchers' anxiety level soared.

By June 1st the fledging window had opened. That was the day the Trostles showed up at the model-boat pond—Glen and Pat Trostle, from Logan, Utah, who had come to New York to show the sights to their kids: Galen was twelve years old and Miranda, seven-and-a-half. The Trostles were attracted to the hawkwatchers for a good reason: They had had red-tailed hawks fledging from a nest in their backyard for three of the last four years.

We exchanged notes about our respective hawks' feeding habits. *Their* red-tails ate mice and voles—textbook buteos. They were surprised to hear ours ate pigeons. Then we quickly moved the conversation to our subject of pressing interest. How did their babies fledge? Their description added fuel to the flames of hawkwatcher anxiety.

"The babies never seem to want to go," said Glen Trostle, "and so the parents entice them out by constant calling. A parent leads a baby to a lower branch, and then, as soon as the chick hops to where the parent is, the parent flies off to another branch. And so on, until the babies get the hang of it." Textbook branchers. How would our hawklets make it across the avenue and into the park?

A fledgling vigil was organized, with volunteers scheduled to be at the hawk bench from 6:00 a.m. to 7:00 p.m. in case disaster struck. Charles Kennedy usually took the sunrise slot, arriving by bike from his Soho loft. He brought a blanket for throwing over a fallen fledgling and a whistle for stopping traffic.

Crowds began to gather at the model-boat pond as the time for fledging approached. The stalwarts arrived at sunrise, their ranks swelled as the day went on, and by evening hundreds would be lining up at the numerous telescopes, hoping to see a hawk baby fledge.

As at all New York events, the TV news folks came around. ABC News managed to get their cameras on a terrace of the building just south of the nest building, the one hawkwatchers referred to as the Ugly White Condo. It looked directly into the nest, allowing the cameraman to get remarkable close-up shots of the young hawks jumping up and down and running from side to side.

A Fledging Pool was organized, and everyone threw two dollars into the pot to buy a two-hour slot. Sharon Freedman signed up for June 2nd, from noon to 2:00 p.m. It was her birthday, and she was determined to win. At 10:30 that morning, as the largest of the babies was flapping especially hard, she called out: "Don't go now! Too early." The bird obediently quieted down and settled in for a rest. In fact, nobody fledged that day, or the next.

Now the hawkwatchers began to wonder if those babies would *ever* leave. Various theories were proposed about what would finally get them going: (1) hunger (if the parents stopped feeding them); (2) the stench of leftover prey and the misery of parasites in the nest; (3) the right weather conditions, with the wind blowing from the east to help them across the street; (4) a surge of courage; (5) the joy of flying. Perhaps it was a combination of all of these that finally did it.

It happened on June 4th, a Sunday. That morning at church Barbara Ascher said a prayer for the baby hawks (and for the hawkwatchers too). She didn't know that the first bird had already flown by the time she got there, but it may have helped retroactively—who knows how prayers work.

It began:

Dear God, Thank you for hawks
That have made us more than we were.
Thank you for opening our hearts.
As their shells opened, so did ours.

And it ended:

Help us to accept, although we pray that they find safe
perches.
And God, we pray especially this night for the first to fly. . . .

TOM FIORE, to nobody's surprise, was the only one there when the first of the three nestlings took off. Something had told him this would be the day, and he decided to get to the model-boat pond even earlier than usual. He got there at 5:35 a.m., just around sunrise. He reported:

When I arrived at the boat pond, all three nestlings were up. They were all facing front and sitting perfectly still. Anthropomorphically speaking, they were anticipating a meal. Maybe they were looking for the other hawkwatchers—Where's Charles? After five minutes they started flapping and jumping. They seemed to be alternating, first one, then another jumped and flapped. This went on for about fifteen minutes. Then the one on the north side started flapping again. He lifted up and I thought, "Uh-oh, this is a pretty good hop." And then he kept on going! He looked a little awkward but he didn't go down. He just flew straight north. He flapped right by Woody's. Then he got to the next building, the one with the big green shade, and landed on the ledge at the top. He folded his wings and looked down and around, as if to say: "Uh-oh, what do I do now?"

I arrived at 6:00 a.m. Three minutes too late! Charles Kennedy arrived two minutes later. Damn! Still, our disappointment was mitigated by relief at the fledgling's success and admiration for his bravery. We could see him there on the Green Shade Building and agreed that the bird was almost certainly a male. For one thing, he looked small, and besides, the first to leave are generally the males, according to Palmer. We watched fledger #1 practice flying skills by taking short, flying jumps from one level of the building to another, maneuvering the landings with his wings. It was as if he were in a tree hopping from branch to branch just like the Trostles' fledglings in Logan, Utah. An enterprising urban hawk.

❧

BIRDWATCHING IS A SPORT, a hobby, a skilled occupation. Hawkwatching is an obsession. Like love, it exhilarates. Like love, it brings anxiety. Birdwatchers watch and listen, ever in hope of something exciting just around the corner. Hawkwatchers exult and despair.

By 5:00 a.m. on June 5th, a small crowd had already gathered at the hawk bench, hoping to see the next baby take off. Obviously nothing was going to happen—Tom Fiore wasn't there, someone noted jealously. Marcia Lowe had taken half a Valium to quell anxiety. Patricia Miller, another devoted hawkwatcher, said she'd been up all night worrying about the fledgling's first night out on his own. She had circles under her eyes.

Anxiety turned into hysteria a little after noon. Fledger, as yesterday's hero had been named, had spent most of the day on top of Woody's water tower, and to the hawkwatchers' concerned eyes he didn't seem happy. "Something's the matter with him. He doesn't look right," said Holly. "He hasn't eaten all day. I think he's starving," said Marcia. Blanche was weeping quietly.

When a large group of tourists on a Suzi's Specialty Tour passed by the hawk bench, Merrill asked if they wanted to see a baby hawk. Thirty-three visitors from Wisconsin lined up and looked at Fledger on the water tower. Then they lined up again to look at the two babies jumping up and down in the nest. They quickly absorbed some of our anxiety and wanted to stay and see if one of the parents came with food for the baby. But Suzi (or a reasonable facsimile) firmly herded them off to their next destination: the Dakota, where *Rosemary's Baby* was filmed and John Lennon was assassinated.

By midafternoon, to everyone's relief, Fledger left his perch on the water tower and began to check out the neighborhood. First he flew to the Ugly White Condo, then to the roof of the nest building itself, and finally to an air conditioner on the Octagonal Building at 75th and Fifth.

Red-tails glide and soar: flap-flap - g l i d e - g l i d e. But Fledger flew like a crow on his first day out of the nest: flap-flap-flap-flap. Mostly, however, he didn't fly at all: He waddled or sidled along the edges of each building he landed on. Sometimes he hopped up and down as he had done in the nest. But out in the world, instead of landing back in the same place, he gave an extra wing push and landed on the next floor, ending up awkwardly balanced on a railing, or planter, or ornamental statue on someone's well-appointed terrace. At each new location he behaved like young creatures generally do: He poked around and explored. Sometimes he simply collapsed in a little heap and had a nap, whether on a ledge or a windowsill.

Around 4:00 p.m., as Fledger perched on top of the Octagonal Building, his father suddenly appeared in the air directly above him and proceeded to demonstrate the art of flying like a red-tailed hawk. Elegantly, the pale bird soared and circled, first toward the Metropolitan Museum of Art where some of the choicest pigeons hang out, then back past the Octagonal Building where the fledgling was watching with raptorly attention, then on toward 72nd Street, then to Pilgrim Hill, where some of the tastiest rats may be found. Finally, he flew back to the Octagonal Building, circled Fledger two more times, and concluded the lesson, landing on the balcony just under the nest. There we could see the remaining two baby hawks in the nest, quietly watching.

Fledger got the idea right away. Hardly had his old man settled down on the nest building when he took off and flew an entire city block to Woody's building without a single flap. All glide. Once again the crowd at the boat pond cheered. "Bellissimo," said an Italian tourist. "Unglaublich," uttered a German visitor—unbelievable.

So far the young hawk had spent his entire life in, or rather on, apartment houses. Unlike most New Yorkers, he had yet to experience the delicious relief Central Park offers the city dweller. On the afternoon of his third day in the wide world, a few minutes before 4:00 p.m., Fledger left his perch on Woody's water tower and sailed into the park.

Jay Sharff, another devoted hawkwatcher, and Marcia had been at the hawk bench since sunrise—they were running on nervous energy. Now they screamed as they saw the bird disappear into the park. They raced from the hawk bench past the Alice in Wonderland statue, following the sound of shrieking blue jays—an excellent indication of a hawk in the neighborhood.

They found Fledger in a red oak. He was sitting near the

top of the tree, swaying a bit in the breeze. It was surely the first time in his life that his perch moved under his feet and he seemed puzzled. His life as a tree-dwelling red-tailed hawk had just begun.

ON JUNE 6TH AT 6:08 A.M., with the usual crowd of morning observers in attendance, the second fledgling took off. Amid the hawkwatchers' excited cries, the bird flew toward the Green Shade Building, headed for the windowsill of the top-floor apartment (the one with the green shade), couldn't figure out how to land, fumbled around a few other windows, and finally flew toward the park, disappearing below the tree line.

"Oh my God, he crashed on Fifth Avenue," someone shouted. Everyone rushed toward the 76th Street park entrance and out onto Fifth. Charles was prepared to stop traffic. For just such a contingency he had tucked a red T-shirt in his backpack. It would serve as a flag. His traffic-cop whistle was in his pocket.

We looked up and down Fifth Avenue—no bird in sight. Then we heard a sound that instantly lifted our spirits: loud blue jay clamor in the park. Back into the park we raced, heading in the direction of the hopeful sound. There, in a pin oak tree just one tree east of Fledger's landing site the day before, we saw the second fledgling. We recognized the baby hawk immediately. It was the one with the reddish chest. Somebody began to call it Lucy—after the redhead of TV fame, I suppose—and the name stuck, though Lucy was quite probably a male.

There must have been a blue jay nest in Lucy's landing tree—the jays were going berserk. They were dive-bombing the fledgling and he did not look happy. He had just left the perfect protection of the nest, and now this! Then, to our deep delight . . . Pale Male to the rescue. For the last hour he had been perched on the top railing of the Octagonal Building, supervising. Seeming to be in no hurry, he floated off and sailed, one might say he aerially sauntered toward the tree where the fledgling was sitting, all hunched up, trying to deflect the blue jay blows. Lucy's father flew by the tree once, then turned and flew by again in the other direction. He was flying low, at Baby's eye level. As he passed Lucy's pin oak, he uttered a cry we had little trouble understanding. "Kreeee-uuur," we heard. "You did good, kid. Don't worry about those jays. They can't really hurt you. Just hang in there and I'll bring lunch soon."

❧

BY THE EVENING OF June 6th each of the first two fledglings had returned from the park to the more familiar precincts of man-made structures. With the warm glow of the sun in the west illuminating the Fifth Avenue skyline, one young hawk could be seen perched on a balcony of the apartment just under Woody's, and the other on the roof of the Octagonal Building. Though they were out of earshot, through binoculars the watchers at the hawk bench could plainly see the fledglings' beaks opening and closing regularly: They were crying, just like babies.

Two out—one more to go. Just before noon on June 7th, Jane Koryn witnessed the last fledging. An addicted hawk-watcher who lived in Woody's building, during the entire fledging vigil Jane took it upon herself to bring trays of buttery Danish pastries and a large thermos of fresh coffee to the early morning contingent at the hawk bench. I remember arriving very early one morning and watching for the light to go on in Jane's kitchen. Coffee coming soon. From her twelfth-floor living-room window Jane could easily see the balconies of the nest apartment. She could see the nest too, but only by leaning way, way out of her window. One fact is clear from her report: She wasn't getting much sleep those days. As she described it to me shortly after the events:

> Last night I looked out after dinner and Mom was sitting on the balcony at the north end of the hawk apartment. I checked at 10:30 p.m. and then again at 1:30 a.m.—she was still there. At 6:00 this morning as I was heading for the hawk bench she was still in the same spot. She didn't leave for four and a half hours.
>
> Pale Male brought something to eat into the nest at 10:30 a.m. A few minutes later I ran to the window because I heard hawk sounds. I could see a lot of activity. Both parents flew by the nest a couple of times, and then circled nearby. Then both the previously fledged babies flew by. I could hear the one that was still in the nest calling—it was the one that still had a few white fuzzy feathers on its head. Then I heard one of the brothers or sisters call as it flew by the nest: Come join us! This is fun! For around fifteen minutes all four—Mom and Dad and the two fledglings—were flying around. Then the parent hawks perched on the black smokestack and had a powwow.
>
> There was a wonderful breeze for the next half hour. It finally happened a few minutes before noon. When the last baby left it flew perfectly, as if it were an adult. It flew right into the park.
>
> And now the reality is sinking in. I still can't believe it. We have an empty nest.

Everyone turned out for the Fledge party—hawkwatchers, Regulars, Fifth Avenue neighbors, hangers-out at the model-boat pond, park employees, park characters—hawk lovers all.

As the party began, the thought of an end ever coming to their fellowship of hawkwatchers made everyone gloomy. That's when Dorothy Poole offered to lead a weekly walk in the Ramble for those who wanted to learn about birds and trees and flowers and butterflies, a walk in all seasons to follow the natural cycles in Central Park. It would have to be early in the morning, for Dorothy had a job, and so, indeed, did some of the hawkwatchers enthusiastic about the idea. It was the beginning of the Earlybirds, a birdwatching group that goes on to this day.

As ever at bird celebrations everyone brought an offering of food or drink, and as the goodies were unpacked, spirits lifted considerably. Charles Kennedy brought a hazelnut cake from the Cupcake Café, his favorite bakery. It bore an icing picture of a noble hawk with outstretched wings and a bright red tail. Though no red-tailed hawk in history had ever had a tail that red—it was the brilliant red of a ripe tomato—the portrait was universally admired. But only briefly; the delicate, buttery cake disappeared quickly.

Jane Koryn brought cookies from a famous bakery on Madison Avenue. Regina Alvarez, the zone gardener for the model-boat pond area, baked a corn cake. Tom brought apples and grapes. Dorothy Poole brought Hawk Crest wine. The Girards and Mo and Sylvia brought home-made appetizers. Norma brought paper cups, plates, and cider.

For the climactic toast Blanche Williamson brought out two bottles of Veuve Clicquot-Ponsardin—the real thing! Weren't we as classy as the hawk building people any day?

SCENE SIX

Hawk-Gazing

❧

All that spring, as the hawkwatchers eagerly followed the progress of the young hawks in the Fifth Avenue nest, the mystery remained. Whenever the afternoon sun hit the right spot at the right moment—as Mom happened to be perched on the edge of the nest feeding the young or on a nearby ledge or railing keeping guard while Dad was out hunting—we'd catch sight of a shiny flash in our binoculars: the band glinting in the sunlight. But even the most powerful birding telescopes could not make out the tiny numbers; the distance between the hawk bench and the nest was too great. As the time for fledging approached, we knew the young hawks would soon disperse and Mom would be harder to locate. Time to go hawk-gazing.

❧

ONE EVENING in the middle of May, on a sudden impulse born of desperation, I approached a young man standing beside an enormous telescope on the lawn outside the Hayden Planetarium and asked: "Could this telescope read some tiny numbers on a narrow aluminum band attached to the right ankle of a hawk perched about two city blocks away?"

Michael O'Gara, the owner of the telescope, thought it was one of the strangest questions he had ever heard. He was there to give some city kids a star show under the aegis of the Amateur Astronomers Association of New York, and he knew he could find Mars for them, and the Beehive Cluster, and the double stars Alcor and Mizar. But birdwatching with an astronomical telescope was something new. Nevertheless, or perhaps for that very reason, the project appealed to him. "How about next Saturday?" was his response.

THE MODERN PRACTICE of bird banding may have originated with the great Audubon himself, who in 1804 attached silver threads to the legs of nesting phoebes to find out if the same birds came back the following year. (They did.) Since then, scientists have been solving some of the most intriguing mysteries of avian wildlife—among them how long birds live, how fast they travel, where they go and by what routes, and whether they keep the same mates year after year—through the recovery of birds with bands.

Experienced researchers with government permits trap birds in near-invisible nets of silk or nylon. They untangle the captive birds, band them on the spot, and then release them unharmed. In addition, a few bird rehabilitators who have qualified for permits—Len Soucy is one—band the birds they return to the wild. In either case, the details of where and when each bird is banded, together with its number, is sent to the Bird Banding Laboratory, a central clearinghouse of information in Laurel, Maryland.

Banding birds is the easy part. The chances of ever coming across a banded bird again are slim indeed. There are simply too many birds in the world, and too few banders. Of the

1.1 million birds banded in North America each year, hardly 5 percent are ever seen again. Most of these are recaptured in some other researcher's mist net. Occasionally a banded bird is found dead and the number reported to the Bird Banding Lab.

Birdwatchers, on rare occasions, do spot a banded bird in their binoculars, as the four Central Park birders did on the October day they first saw the hawk with the silvery band—the one who went on to become Pale Male's mate and proud mother of three. But the obstacles in the way of reading eight or nine tiny numbers engraved on a band fastened to the ankle of a living, moving, flying bird are formidable, even when the bird is as large as a hawk.

Hope now centered on Michael O'Gara and his Newtonian Reflector, a powerful telescope he had constructed himself. It could magnify 237 times. It could perfectly reveal the four moons of Jupiter, and the rings of Saturn. Perhaps it could read the numbers on Mom's band.

MAY 26TH WAS THE DAY Michael O'Gara first brought his formerly stargazing now band-reading telescope into the park. The hawk chicks were no longer babies. They were almost full-grown and beginning to jump up and down in the nest, practicing for their first flight.

The park, too, had passed its annual newborn-baby stage; a darker green prevailed. The black locusts were still in bloom, their white flowers so fragrant you could smell it a block away; the black cherry flowers were beginning to fade, and the horse chestnut blossoms had reached their peak two weeks earlier. Lindens, the last trees to bloom in Central Park, were just coming into bud that week.

Flowering trees belong to spring, but as Michael O'Gara wheeled a luggage carrier bearing his huge telescope and accoutrements down the hill from 72nd Street to the hawk bench, summer was clearly on its way: The blackpoll warblers had arrived. Small grayish birds with a black cap and white cheeks, they are always the last spring migrants to arrive in the park, perhaps because their journey is one of the longest; they winter as far south as Brazil, while they breed in the spruce-fir forests of northern Canada and Alaska. Rather unbelievably, some of these tiny birds manage a nonstop flight of 2,500 miles over the Atlantic Ocean.

A couple of blackpolls were singing their high-pitched, thin, mechanical song in the elm above the hawk bench—Zri-zri-zri-zri-zri, getting louder and then softer, like a faulty generator with some loose part rubbing against some other. The weather was gray and forbidding, with rain predicted for the late afternoon. Not a propitious forecast for the task at hand: to make out eight numbers, each about the size of a digit on a standard push-button phone, from a distance of about a third of a mile away. (Charles Kennedy did some elaborate measuring and figured it was 560 yards from the hawk bench to the nest.)

I carried a certain number on a slip of paper in my handbag—1387-38569, the band number of First Love. It was always with me, just in case. I knew it was a long shot, but still I was hopeful.

Charles was infatuated with this possibility. Norma, too, was rooting for First Love. Anne Shanahan said, "Wouldn't that be wonderful?" though in her heart of hearts she still thought the bird was Chocolate.

Both Charles and Norma were there by noon on May 26th; Sharon, Anne Shanahan and Tom Fiore arrived shortly thereafter. The new band of hard-core hawkwatchers was there too: Merrill, Marcia, Holly, Blanche, and Jim.

In spite of the weather there were still great numbers of people going by—tourists, dog walkers, art lovers heading for the Metropolitan Museum, and local residents out for a Saturday walk in the park, rain or shine. As Michael O'Gara set up his huge telescope at the edge of the pond, a good distance from any trees that might drop sap on the valuable instrument, a crowd began to gather at once.

O'Gara, an obliging man, let everyone have a look. It was 2:30 by the time he could begin serious monitoring of the nest. Mom had just arrived with some prey, but she was facing the wrong way. It was after 3:00 p.m. by the time she turned and showed the leg with the band. The sun was shining directly into the nest, and without taking his eye from the eyepiece, Michael O'Gara said, "I see numbers!"

Numbers!

"I see the crimp and a 1 and a 3, yes, I definitely see them clearly!" O'Gara called out. The crimp, I knew, is the ridge where the band is fastened. It would mark the beginning and end of the number. I sneaked a look at my slip of paper and my stomach jumped with excitement. That number began with a 1 and a 3.

Just then Mom took off. She flew to Woody's building and perched on one of the angel ornaments a few floors below his penthouse. Again, though she was looking to the south, her body was facing north. Only her left foot was visible. She stood there for almost twenty minutes without moving.

While waiting for her to shift positions, the old-time hawkwatchers began reminiscing about the past. "Remember how depressed we were this time last year when the eggs didn't hatch?" someone asked, and we happily remembered. Depression recalled in the midst of elation is oddly uplifting.

Sharon, who was taking a turn at the Newtonian Reflector,

suddenly called out: "I see a 7 and a hyphen." It was 3:45 p.m. Michael O'Gara took over to confirm her sighting. At that moment the hawk shifted on her perch and O'Gara exclaimed: "I see another number after the hyphen. Wait a minute. Yes, it's a 3." We all lined up for a look and all of us saw it: a 7, the hyphen, and then a 3.

"What's the number on the first bird's band?" O'Gara asked.

I took out my slip of paper and handed it to him: 1387-38569. He looked at it and said: "This has *got* to be the same bird! That bird had a 7 hyphen 3, and this one has a 7 hyphen 3. The first bird had a 1 followed by a 3, and this one does too. What are the statistical odds of it being a different bird, with two identical combinations of numbers? That's beyond the realm of coincidence."

Charles and I exchanged a look, did a high five. Yes! It had to be First Love!

ON JUNE 3RD O'Gara brought his telescope down to the model-boat pond again. Once again, as Mom fed her babies, she was facing the wrong way. She landed once on the window ledge just under the nest and once on the building's ornamental cornice, and each time she was facing north. Finally, she landed on Woody's TV antenna with her right leg in perfect sight, but by the time O'Gara turned his telescope to the new location, she had flown off into the park.

A new angle was needed, and Jane Koryn, the coffee and pastry lady who lived in Woody's building, provided it.

"Maybe you'll get a better view from my living room," Jane said to O'Gara when his second attempt at band-reading had yielded no more numbers. Her offer was immediately accepted.

The very next day the first of the three nestlings took off, and within a week all three had fledged. Was it too late to take advantage of Jane's view, now that the babies had left the nest? Yet all that week and the next the newly fledged young continued to hang out on nearby roofs, windows, antennas, window boxes, and even, every so often, on the nest itself. Mom was usually perched somewhere nearby. Jane Koryn's close-up view still seemed well worth a try.

ON JUNE 15TH Michael O'Gara, Charles Kennedy, and I arrived at Jane Koryn's high-rise apartment house on Fifth Avenue just across 74th Street from the hawk building. O'Gara was wheeling his Newtonian Reflector. Charles balanced a camera with a telephoto lens in one hand and a tripod on his shoulder.

The liveried doorman looked at us dubiously as he buzzed the Koryn apartment on the intercom. Though Jane confirmed that we were invited guests, he hastily ushered us into the service elevator. Was it our non-designer clothes, our non-stylist hair, or something about our non-cool expressions of suppressed excitement at the task ahead that gave him the clue? We protested only mildly—it was obvious that employees of Fifth Avenue apartment houses have little experience with birdwatchers or astronomers.

No one spent much time admiring the Koryns' elegant apartment with its panoramic view of Central Park, its antique furniture, Oriental rugs, and artworks. The focus was on hawks, or, rather, on a few inches of a single hawk's right ankle, just above the deadly talons but below the feathery pantaloons that cover its thighs.

A mere fifteen minutes after O'Gara finished setting up his telescope at an open window facing the nest building, the female hawk landed on the ledge of the hawk apartment directly in view. Though the bird was still facing north, from this vantage point the right leg was in clear sight. The bird was so close that the band was visible to the unaided eye.

O'Gara inserted an eyepiece, fiddled with the lens, and worked at the focus. "I'm getting it," he muttered. "I see the band," he said with considerable excitement. At that moment, as if impelled by some malign force, the bird tucked her right leg into the mottled feathers around her belly, and began to preen. Hawks, like most birds, spend a lot of time at feather maintenance, at least 50 percent of their waking day, in fact. This bird now, with her right foot tucked out of sight, proceeded to spend one and a half hours working over what seemed to be every damned feather on her body.

To ease the hawkwatchers' frustration, Jane brought out a bottle of Pinot Grigio, four exquisite crystal glasses, one bowl of apricots and another of fancy almonds. As we downed the spirits, our own lifted.

A little before 5:00 p.m., perhaps assisted by the Pinot Grigio, O'Gara began squeaking like a mouse. The hawk looked over, but the leg remained tucked. O'Gara sang a few lines from a Schubert lied—now a video producer, he had once trained as a baritone—but the bird did not bat an eye. "Hey, you on da ledge! Down! Both feet down!" he called out in a Dead End Kid voice. No response. Not until 5:55 p.m. did the bird put her right foot down. But she also scrunched up and the band was obscured by feathers.

At 6:35, as the light began to fade, Mom straightened up and showed us her band. O'Gara had the telescope perfectly focused this time, and he called out: "I got it! 1387 hyphen 3!"

It was virtually certain. There were now five identical digits to those on First Love's band. But O'Gara continued to peer into his telescope. Then he called out: "I see a 2, a definite 2."

We looked at the slip of paper with the numbers, and then at each other.

"Are you sure?" Charles asked anxiously, hoping the answer would be no. For there was no 2 in the first hawk's band number.

"I'm one hundred percent sure," O'Gara answered. "Wait, she's moving her foot, I can see more numbers. Yes, it's 38627 and then the crimp."

We sat there without speaking for a long few moments. It was hard to believe. Putting everything together, we had a complete number now, 1387-38627, which differed from the band number of First Love by only three digits. Finally, Charles broke the silence.

"The end of Romance," he said.

❧

THE BIRD BANDING LABORATORY is where all bird-banding records are stored. Until recently it was part of the U.S. Fish and Wildlife Service; now it is part of the U.S. Geological Survey, Biological Resources Division. The information gleaned from retrieved bands is mainly used by scientists. But if members of the public happen upon a banded bird and transmit the number to the Bird Banding Lab, they will be told where, when, and why the bird was banded. They will also receive a certificate suitable for framing with information about their banded bird.

Mary Gustavson, a biologist at the Bird Banding Lab, took my call. First she described how the band numbers work.

"A band is composed of three or four numbers before a hyphen, and then five numbers. The numbers in front of the hyphen are the prefix," she explained. "And the last number of the prefix indicates the size of the band. The number you just gave me—1387-38627—has a 7 in front of the hyphen. That's a big size, used for banding red-tailed hawks, for instance, or snow geese, or mallards.

"A 1387 prefix also indicates that it's a lock-on band. Most bands are what we call butt-end types, but a hawk could grab a butt-end band and pull it off. That's why raptors have to have lock-end bands. The two ends come together, one end has a short tab, and that's folded over, locking the band on the bird's leg firmly."

I didn't want to hurry her. But I could hardly wait until she looked up my number and gave me information about our banded female. Unfortunately, that would take her a while. She promised to call me back within a few hours.

She did. Finally I had an answer, though one that led to even more baffling questions.

"The bird you called about—1387-38627—is an adult redtail at least two full calendar years old, rehabilitated by Len Soucy at his center in New Jersey and released there last October 8th.

"That's incredible!" I exclaimed. "The bird we originally thought this was going to be was also a Len Soucy bird, one that he banded three years ago. Is there any chance that he made a mistake recording the number, and this one actually *is* the other bird?"

"Oh no," she answered. She sounded horrified at the very thought. "Banders are required to keep exact records, and when they band a bird, first they read the number on the band, then they put it on the bird, and then they write it down and read it again, to make sure all the numbers are right and in the appropriate order. Then they have to send the records to us."

"But doesn't it seem a bizarre coincidence to have two Central Park red-tails banded by the same bander? And isn't it amazing that the two numbers begin with the same six digits?"

Ms. Gustavson had a good answer to both questions:

"We have only two thousand banders across the country. They're a very scarce commodity, and particularly banders of hawks. Not all banders work with hawks because it takes special skill to be able to band and then release a hawk safely. Len's an active bander and bands all his rehabilitated birds. He probably bands a large percentage of the nonmigrant red-tails within the state. So it doesn't surprise me that both birds would connect to him.

"As for the identical digits: Len Soucy received a big batch of bands with sequential numbers. He had bands numbered from 1387-38501 to 1387-39000. Consequently, the bands of any two hawks rehabilitated by Soucy in recent years would start with the same five or six digits."

A perfectly plausible explanation. It reminded me of certain magic tricks I've seen: As the trick is performed it seems impossible to explain by any earthly means. Then, when I find out how it's done, it seems so obvious, and my earlier certainty that I'd witnessed something supernatural seems so foolish.

"Can you tell me more about bird 1387-38627?" I asked Mary Gustavson. "What happened to her? How did she end up at Soucy's?"

"That's all the information I have," Mary Gustavson said, concluding our conversation. "I'm sure Len can check his records and tell you all the details."

I CALLED LEN SOUCY as soon as I hung up, but I got his answering machine. I left a message, giving the band number and the present status of the bird.

Soucy called me back that afternoon. He sounded excited.

"It's awesome that you succeeded in reading the numbers," he began. "That happens once in a million with a live bird. And wait until I tell you the story of this bird!

"Last year on September 6th, I received a call from the Bergen County Animal Shelter. Someone had brought in a red-tailed hawk that was found injured on the Palisades Interstate Parkway, near Alpine, New Jersey. So we went over and picked the bird up.

"The bird had suffered severe trauma to her right eye. I guessed the bird was a female because of her size.

"She was extremely debilitated and thin when we brought her in, probably as a result of the injury and not being able to hunt. We had to force-feed her for a week to keep her alive.

"We treated the eye injury over a period of time, but after our vet looked at it with an ophthalmoscope, his best judgment was that the eye was nonfunctional. In the five weeks we had her, there was no pupillary contraction or dilation in that eye, no response to light. So that's the story of your hawk: The bird is almost certainly blind in the right eye."

I had to interrupt him. "That would explain why she was always facing north whenever we saw her on a Fifth Avenue building or the nest. She was looking with her good eye."

"Sounds right," he agreed. "We agonize over these one-eyed birds. A lot of other rehabilitators I've talked to believe that one-eyed hawks have almost no chance of surviving in the wild. Hawks are terribly dependent on eyesight for their living. And with one eye they have no depth perception. How can they avoid obstacles—they are literally blindsided.

"That's the awesome part of this story: While vision is so necessary to diurnal birds of prey, this hawk with totally compromised vision is not only making it, but she's reproducing! Three kids! Wait until science hears how she is doing. A lot of people say that one-eyed birds are not viable in the wild."

He paused. "I suppose I shouldn't tell you this, but I was a hair away from euthanizing this bird."

"Why did you end up not doing it?" I asked.

He thought for a moment. "I don't know what made me decide, except that I get so bloody tired of killing birds. When I started as a rehabilitator thirty years ago, my purpose was not to euthanize. I wanted to fix them all and let them all go back into the wild in perfect health. Some of them are beyond fixing, however, though we did release twenty-one hundred birds last year.

"Besides, there was something about the bird itself. It responded so well to our treatment. It started out in such terrible condition, but then it became stronger, more self-sufficient. It could catch a mouse in our flight chamber, so it was self-feeding. All these are signs to me. I hope this isn't anthropomorphic, being a scientist, but this hawk looked like it wanted to live. It was trying to live, gritting it out, if you will, and I said to myself, 'Let it live.' "

Len Soucy kept the bird for a month and two days. On October 8th he banded and released her at his center in Millington, New Jersey. Less than a week later the hawk was in Central Park, eating a fat rat in an oak near Bank Rock Bridge.

SCENE SEVEN

Chocolate's Story

Once the hawkwatchers learned who Mom wasn't, they returned to the question of who she was. A small, unlikely idea had been waiting in the wings, as it were, and now its time had come. Fleshing it out with all the facts they had come to know about red-tailed hawks, their habits and way of life, the hawkwatchers added a healthy dollop of anthropomorphism and came up with a satisfying fantasy:

CHOCOLATE AND PALE MALE always took separate vacations at the end of the nesting season. Pale Male never went far—exploring the northern and southern ends of the park was fun enough. Chocolate liked to range farther. She often wandered as far north as Inwood Park at the northern tip of Manhattan to check out the action. Once she had even spent a week in the Bronx! This time she thought she'd go farther still—she needed a little pick-me-up after all that sitting in the hot sun and nothing happening for the second year in a row. She decided to visit a special place from her childhood.

She headed west. She didn't have to fly all that far, just across the big river to that wooded area a little bit south of the George Washington Bridge. She had been born in a nest near that spot and had a particular fondness for the landscape and the local rodents there.

She had a wonderful trip, catching all the right winds and thermals. Then, just as she was swooping low in pursuit of one of the fat meadow voles she'd been dreaming about (those city rats never had much flavor!), the terrible thing happened. Suddenly something big was bearing down on her, much bigger than her major enemy, the great horned owl. It was moving fast, it was directly in front of her, she swerved to the left, and . . . the next thing she knew she was in the hawk hospital. There she stayed until she was as good as new, except for that funny thing with her eye and the little crick in her wing. Just before the man with the white beard took her out of a box and let her go flying free again, he placed a silvery band on her right ankle.

She zipped back to Central Park as fast as she could.

"I was about to give up hope that you'd ever get back," her pale mate uttered in his particular way, not letting on that he had already started casting his eye around for her replacement. That's what he'd done with his first love, as Chocolate knew all too well, for *she'd* been the replacement then. This time everything worked out fine. Pale Male brought her an extra-fat pigeon, all plucked and ready to eat, to show that his love was still strong.

WHY NOT? ACCORDING to this scenario, the bird with the band that the three birdwatchers had seen on October 14th was not an unknown hawk at all. It was Chocolate, a little bit worse for wear, wearing a fancy new ornament on her right ankle, but still the same bird.

SCENE EIGHT

Hawk Nursery School

On their very first weekend out of the nest, the Fifth Avenue fledglings were faced with three ordeals. *Pocahontas* came first. On June 10th the full-length Disney cartoon was to have its world premiere on the Great Lawn and the park had been in an uproar for weeks. Teams of workmen erected a 60-foot-high stage; a small flock of mechanical cranes installed four 100-foot screens on top of cloth-draped red steel shipping containers. A huge tent was raised for entertaining bigwigs; light fixtures sprang up like giant fungi after a rain. Finally, the entire area was cordoned off with gray metal barriers so that ticket-holders only could attend the event.

There was lots of grumbling about *Pocahontas.* The ballplayers resented the loss of their playing area. Civic-minded citizens were indignant that a public park was being used for a commercial venture. The historically minded cringed at the gross violation of Vaux and Olmsted's vision of the park as a pastoral, peaceful resource for harried city dwellers.

Naturally enough, the park's birdwatchers groused too. They worried about the birds. How would the lights and noise and commotion, to say nothing of the extra pollution from all the infernal machinery, affect the wildlife in the area? Noting that the giant screens and light fixtures and projectors on the

Great Lawn effectively blocked all views of real trees for the audience, birdwatchers were not impressed with statements issued by the Disney publicity department declaring that the film was "about nature, about living in nature," and therefore Central Park was a perfect place for its premiere.

As the day of the opening approached, the hubbub at the Great Lawn verged on pandemonium. Rock music blared all day from the oversized loudspeakers, while a chorus in one place and an orchestra in another (within earshot) rehearsed altogether different music. What about the newly fledged hawks? Surely they'd be spooked by the noise. When someone said there'd be fireworks at the end of the performance, *Pocahontas* took on nightmare proportions.

At 11:00 p.m. on June 10th Tom Fiore heard the fireworks from his apartment on West 82nd Street and wondered how the hawks were doing. Norma Collin, who lives at Second Avenue and 78th Street, five blocks from the park, heard the fireworks too. Indeed, the noise woke her up. "My whole apartment shook for fifteen minutes," she reported. Norma is not a bigtime worrier, but that night she worried about the baby hawks. Even I could hear a dull roar at Riverside Drive and 92nd Street, as far west from the park as Norma's apartment is east, but considerably farther north. These may have been the loudest fireworks ever heard in New York City. They must have been amplified by all those loudspeakers.

The next morning I headed for the hawk bench a little after sunrise. I had tossed and turned all night, full of foreboding. On arrival, instant relief. For there was the entire hawk family in plain sight: one baby on the roof of the Green Shade Building; one on Woody's lower terrace, perched on a planter; and the third on Jane Koryn's window ledge a few floors below. Mom was on the black smokestack and Pale Male was circling around with a pigeon in his talons. He dropped it off on Woody's terrace and went out to hunt for more. The hawks had survived *Pocahontas.*

The next day was the Puerto Rican Day parade. Every year on that occasion thousands of parade-goers carrying flags and six-packs and radios pour into the park at every entrance looking for a place to sit down, eat, drink, sing, dance, and celebrate. Park workers hate parades—the revelers leave a terrible mess. People in the neighborhood hate the noise, not of the parade itself but the horns of infuriated drivers stuck in the traffic jams it creates. Was it the noise, the additional crowds in the park? For whatever reason, the hawks vanished for most of the day.

That same evening Nature put on an extravaganza of her own, with thunderclaps following lightning bolts by mere seconds. The storm was close and loud as hell, easily outstripping the *Pocahontas* fireworks in decibel level. As I worried about the baby hawks, I knew every other hawkwatcher was doing the same.

The next morning I arrived at the hawk bench at 8:00 a.m. to find Charles Kennedy already there. He was beaming. All three fledglings were standing in the nest, just like in the good old days. Fifteen minutes later Mom arrived with a large rat for the fledglings' breakfast and promptly flew off again in the direction of the Ramble.

What foolish worriers we turned out to be. The hawklets were fine, not a feather out of place. The hawkwatchers were nervous wrecks.

❧

THE THREE FLEDGLINGS MOVED into the park for good toward the end of June. Thereafter, from a relocated hawk bench near Cedar Hill the enraptured hawkwatchers, along with thousands of casual passersby, watched Central Park's first hawk babies go from kindergarten to college, as it were, right in full view.

On June 8th Norma Collin saw Pale Male bring a big pigeon he had caught, killed, and plucked to a fenced-in area near Cedar Hill—the Killing Field as the hawkwatchers called it. To be sure, it was not fenced in for the convenience of the hawks; it had recently been cultivated and re-seeded. Still, for a few weeks it offered the hawk children an undisturbed place for feeding, resting, and pursuing various educational activities, chasing ants, for example. When Pale Male arrived that day, all three hawklets were on different branches of a half-dead old oak tree—the Killing Tree—noisily clamoring for dinner.

A few days later Merrill Higgins, who had moved his scope and signs to the new hawk bench, observed a hawk juvenile awkwardly attempting to catch his own pigeon on the Killing Field. "Failed conspicuously," he noted in the Bird Register. Now that the hawks were near at hand rather than high on a building, people came up to Merrill with different kinds of questions. More than one passerby asked: "Is that bird yours?" or "Who owns that bird?" Others asked: "Are those birds dangerous?"

On June 26th, on their lunch break, Holly Holden and Blanche Williamson saw Pale Male bring a live mouse to one of the hawklets, no doubt for the same reason that cheetahs on the plains of Serengeti bring live Thomson's gazelles to their growing young: to teach them how to hunt. The red-tail youngster went after the mouse, but it scurried away—close, but no cigar.

On June 27th Anne Shanahan saw the smallest fledgling catch what may have been his first rat just outside the Three

Bears playground at Fifth Avenue and 76th Street. That particular playground was one of the hawklets' favorite hangouts for a few weeks. They liked to perch on the fence at its northeast corner, near the sprinkler, while toddlers splashed and played under the idle supervision of a parent, a baby-sitter, or an occasional Mary Poppins-like British nanny, sensible shoes, hat, and all. Neither the children nor the caretakers showed the slightest fear of the birds of prey. Most of the time they failed to notice them, though the birds were often perched no more than a few feet away.

All through June the young hawks continued to beg for food from various perches in the area, producing pathetic gull-like sounds for hours at a time—Kleek, kleek, kleek. This made them easy to find, even when perched deep in a leafy tree. Though they were now as big as their parents, there was no mistaking them, even when their telltale tails—brown instead of red—were not in view: the babies looked clean and fresh and bright, their feathers in perfect condition. They gleamed. Pale Male and Mom, in the meanwhile, looked worn-out. Their feathers were frazzled, and each had missing primaries as they began their annual molt. A tough job, feeding three kids and teaching them the ways of the world.

By the second week of July the hawk family had moved operations to the north. The Obelisk, also known as Cleopatra's Needle, then became a favorite hawk (and hawk-watcher) gathering place. On July 19th Anne Shanahan saw Nester, as the last to fledge was known, catch a sparrow in the shrubbery near the base of the Egyptian monument. He plucked a few feathers, and then ate everything else—head, bones, and all.

On July 21st, after a rain shower, Norma saw all three young hawks drinking and bathing in a puddle just behind the Metropolitan Museum. On the 25th Anne Shanahan saw

Lucy—everyone now agreed about his sex—catch a pigeon on the sidewalk of the 86th Street transverse. He took it to the wall at the road's edge and devoured it while trucks, cars, buses, and a screaming fire engine zipped by. Finally he cached part of it in a nearby tree and flew into a playground adjacent to the road. He landed on a pyramid-shaped climbing apparatus and sat there for a long time without moving, looking for all the world like Horus, the falcon-headed god of ancient Egypt.

There were rare sightings of the fledglings all summer, including one on the office window of Paul Sweet, a research assistant in ornithology at the American Museum of Natural History and an ardent Central Park birder. As the summer wore on, the three young hawks were around but elusive. They seemed to be losing their fearlessness toward humans, a good thing for them, according to Len Soucy, who often declares that the greatest danger to urban red-tails comes from humankind.

The parents, too, were hard to find. Pale Male was seen only once or twice that summer. The last sighting of Mom had been on July 23rd; no sight of her at all in August.

By the end of July the handwriting was on the wall—the hawk show was closing. For the hawkwatchers one consolation alone kept depression at bay: the whole grand spectacle was scheduled to re-open next spring, same time, same place, same glorious cast of characters. But we had overlooked the great rule that governs human and hawk affairs alike: things change.

ACT IV

The Queen Is Dead, Long Live the Queen

If any part of nature excites our pity,
it is for ourselves we grieve . . .

THOREAU'S *Journal*

SCENE ONE

Jerry Domino's Letter

Jerry Domino, an engineer in New Jersey, was clearing his desk one December Sunday when he came upon an issue of *Smithsonian* magazine that had been sitting there unread for almost two months. As he idly leafed through it, a blurb on the contents page suddenly caught his eye: "Birdwatchers set out to solve the mystery of the banded hawk." He had a special reason to be interested in the subject of bird bands and immediately turned to the article. Even before he finished reading he took a small object out of the desk drawer and examined it carefully, comparing it to information in the article. More than a little excited now, he read to the end and then promptly wrote a letter to *Smithsonian*.

I had written that article, telling the latest act of the Fifth Avenue Hawk drama. Three days later Suzanne Crawford, my editor at the magazine, telephoned me. "I have some upsetting news for you," she began. My heart sank. Had I made some embarrassing factual error that had just come to light?

"Let me read you a letter we received," she continued, "We've checked it out and believe it is accurate." She proceeded to read me the following letter, which would appear in their next issue:

Dear Sir:

With growing interest I read Marie Winn's article about the nesting red-tailed hawks of Fifth Avenue in New York City. I am a surveyor with the New Jersey Department of Transportation.

While working on the Palisades Interstate Parkway, in late September, I discovered a dead hawk with an aluminum ankle band about five miles north of the George Washington Bridge and a few hundred feet west of the palisade. A co-worker knew the importance of the band and removed it, giving it to me to pass along to the proper authorities.

I regret to inform Marie Winn and the Central Park bird-watchers that bird No. 1387-38627 ("Mom") is dead. Its location, just behind the guide rail, led me to believe it had been hit by a car.

Jerry Domino
Metuchen, New Jersey

"It's impossible!" I remember saying as soon as Suzanne finished. "I just saw both of the hawks near the nest today. They're adding sticks and getting ready to start a family again! Mom couldn't possibly be dead."

"I'm afraid it's true," Suzanne answered. "It must be a different hawk up there. I'm so sorry about this."

"Are you absolutely sure?" I asked. It didn't make sense. But Suzanne seemed convinced. "Domino promised to send the band to Len Soucy. That will be conclusive," she said.

❧

"THOUGH IT BE HONEST, it is never good to bring bad news." Would that I had taken Shakespeare's good advice. Instead, I immediately set out to pass my bad news along to the hawkwatchers. Norma wasn't home. I got Charles' and Tom Fiore's answering machines. Then I reached Anne Shanahan and told her about Mom.

After a long silence, Anne's first words were, "I hate this story!" I could hear she was upset. Belatedly I realized I should have cushioned the blow.

"I'm glad you're telling me," she hurried to assure me in her ever-kindly way, "I know this is life, and it's not that I don't want to know the truth. . . ." There was another long pause. Then she asked, "Has anybody seen this band?"

"No, Domino promised he'd send it to Len Soucy," I told her.

"Well, I'm not totally convinced. We don't really have any evidence except this letter. I think we should wait until we hear that Soucy has the band. This man might be making the whole thing up."

"Why should he do such a crazy thing? And besides, he read the number to Soucy, and Soucy checked it against his records," I told Anne.

"But the number appeared in your article!" Anne exclaimed. "That's where he got it, not necessarily from the band. He could easily be making the whole thing up."

She was right. "I guess we'll only know for sure when Soucy gets the band from Domino."

The next evening Anne Shanahan telephoned. "I've been going through my notes from last fall," she said, "and if I saw what I saw, something doesn't sound right."

She proceeded to read me her notes of October 16th, which I took down carefully:

> Female hawk on the pipe ["That's the big building on Fifth Avenue and 78th Street, the one with the black smokestack," she explained]. On concrete bridge east of the hawk tree I saw

the two hawks flying quite low over me, flying together and giving the scream. The female hovered for a minute and then they circled together. Twenty minutes later they're in the pin oak in the field west of the Boathouse parking lot. I heard the jays and then the squirrels making the bleating sound. One squirrel ran into a hole on the side of the tree. I looked up to see ("This is the thing I was looking for!" she interrupted her reading to say) the female hawk. *Also saw her leg band.*

"Wow," I said. October 16th was well after Domino claimed to have found the dead hawk by the highway. If Anne Shanahan saw a banded female on the 16th, it seemed almost impossible that it wasn't Mom.

"Now, I may be out to lunch," Anne said, "maybe I didn't see it. But I thought I did, because I wrote it down here. I'm just reading you what I wrote."

I certainly didn't think Anne Shanahan was out to lunch. She is the most dependable observer in the park. I called Suzanne Crawford at *Smithsonian* to suggest that they hold off publishing Domino's letter until Soucy had the actual band in hand. But the January issue had gone to press. Besides, she sounded dubious about what I was telling her. I had a feeling she thought I was simply denying reality.

A few days later I received a call from Merrill Higgins.

"When I heard this story about Mom being dead, I was devastated," he said. "But then I began to get skeptical. I have photos of her taken in the spring, and some I took a few weeks ago, and she doesn't look any different. I was satisfied with the color. That's the reason I brought the scope in on Sunday. I was determined to find the female.

"First I saw a hawk on the nest apartment window railing just beneath the nest and I put the scope on it. I could see it was Pale Male. I went to look for the female in some of her favorite spots and I was gone about fifteen minutes. When I came back to the bench, there was still a bird in the same window. But lo and behold, it was the female. They had traded places. Another indication that it's his regular mate.

"I put the scope on her and couldn't see anything, not even her tail, which was behind the grille. Then she turned."

Merrill paused for dramatic effect. "I clearly saw the band," he said emphatically.

"I saw the band," he repeated. "Norma saw it too. I wanted a witness and I got her to look in the scope. That bird's not dead."

That sewed it up. It was a hoax, it had to be. Domino must be completely demented.

I waited eagerly to hear from Len Soucy and three weeks later the call came: "The band from Domino arrived today, and the number was 1387-38627. The bird he found was Mama. That's the news. I'm sorry."

Well, that's that, I thought. But it wasn't.

SCENE TWO

Romance Romance Romance

It was one of the richest owl seasons in Central Park history. Instead of the usual two or three sightings, that year there were more than thirty.

A pair of long-eared owls had arrived in December and roosted in a blue spruce at the bottom of Cedar Hill for more than three months. From late October to the middle of March a procession of saw-whet owls popped in and out of various hiding places throughout the park—at least seven sighted on one unbelievable day. Reading from north to south: one just south of the Harlem Meer, one in an evergreen grove at the north end of the Great Lawn known as the Pinetum, four in the Ramble, and one in a cedar just outside the zoo cafeteria around 64th Street. (That December a saw-whet was spotted on a Christmas tree being sold at Broadway and 86th Street; a month later another actually flew *into* Charlie's Gourmet, a small grocery store at 68th Street and Columbus Avenue, causing a near riot.) Clearly it was one of those cyclical "irruption years" that occur once a decade or so, usually as a result of food source declines up north.

Even as the owls invaded the park, Pale Male and Mom—or whoever she was, the great, beautiful bird with the dark head and the bright russet tail who looked the spitting image of Mom—were getting ready for a new season: twig-gathering, courtship flights, mate-feeding. On March 5th Anne Shanahan and Marcia Lowe wrote an entry in the Bird Register beginning with the word "CONGRATULATIONS." On a window rail on the next-to-the-top floor of a building on Fifth Avenue and 73rd Street, at 1:50 that afternoon, the hawkwatchers saw what they had been eagerly awaiting: The female hawk fluttered her slightly outstretched wings in invitation. Pale Male hopped on her back and they mated.

On March 23rd, just as the Fifth Avenue hawks began to sit on their eggs, the hard-core hawkwatchers resumed their vigil at the hawk bench. They were not buying the Mom-is-dead story. Of course they knew about Jerry Domino's letter—someone had cut it out of *Smithsonian* and stuck it in the Bird Register. But they didn't believe it. Hadn't Merrill and Norma seen a band on the right ankle of Pale Male's present mate? How could there be *another* banded female?

Even after I reported that Len Soucy had received the actual band from Jerry Domino, doubt prevailed.

"Do you think Len Soucy is not telling the truth about the number on the band?" I asked Holly Holden, who adamantly rejected the idea that last year's successful mom had been found dead.

"No, I don't think he's lying. But maybe you guys read the wrong number with the telescope last year. Those numbers are so small. It's easy to make a mistake."

If only the whole unpleasant story could be scotched once and for all! If only they could *prove* that Mom was alive and well. There was just one thing to do: read the band attached to the ankle of the female hawk now sitting on eggs at the Fifth Avenue nest.

❧

MERRILL HIGGINS had spent the whole spring giving people looks at the hawks through his telescope. It was only fair and proper that he should be the one to look through the same telescope and make the great discovery. He told me the story in full detail a few days after it happened:

"It was last Sunday, March 31st, and I was sitting at the pond that afternoon with Holly, Blanche, and Marcia. The female had been sitting on the nest all day, and finally, around five-thirty, Pale Male arrived with a pigeon. She took off with it, and he settled down on the eggs.

"So the three girls ran to see if the female took her pigeon to the usual tree she liked to eat in, the one near Summer-Stage. I was tired and hungry—I'd been there all day and I was ready to go home. Suddenly Holly comes running back. 'You've got to bring your scope. She's on the tree with her pigeon and you can see the band!'

"Well, I didn't really want to go. A few weeks earlier we had been really close to the female—no more than fifty feet—and we could see the band clearly without binoculars. But when I focused my scope on it that time I couldn't see any of the numbers. It seemed impossible. Obviously we needed the astronomy telescope to read the numbers.

"I really thought I was wasting my time, but they all insisted so I dragged my scope up the hill from the model-boat pond. I didn't even bother to climb the last little bit that would have brought me as close to her as possible. I just set up the scope on the East Drive. I was about twice as far from her as I had been that other time, when I hadn't been able to read the numbers.

"I framed the image in my scope. And I couldn't believe it! I could see the numbers clearly this time. It was as if they were lit up like a neon light. It's because the sun was low on the horizon this time—that's what made the difference. That other time I couldn't read the numbers it was early afternoon and the sun was high.

"I could read the numbers 569 right away. Then I saw a 3 and an 8 just before the 569. That gave us 38569. And then I saw the 1387 and the hyphen. I wanted Holly, Blanche, and Marcia to see the numbers too because I knew they'd never believe me unless they saw it with their own eyes.

"I waved and pointed and they all came running. And all of them saw the number. Marcia immediately said: 'That number sounds familiar.' And she was right. It was the number of Pale Male's first love, the year he first came into the park."

WHEN I CALLED to tell Charles the news, he gave a volcanic whoop: "Yahoo! His first lady's back in her rightful place at last!"

He paused. Perhaps it felt unseemly to rejoice when Merrill's sighting confirmed the sad reality about last year's hawk mom. Then Charles found reason to exult anew: "The Queen is dead. Long live the Queen!" he exclaimed. "Romance, romance, romance!"

PHONE INTERVIEW WITH Len Soucy:

"So she's back in Central Park? That's awesome. I've banded thousands and thousands of birds. The probability of this being the same bird I banded four years ago—independent of other factors—would be a million to one, well, a thousand to one for sure. Things like this just fracture me. We think we know everything, but we really know little.

"One thing we do know: even though the pair bond of these birds is strong, once one of a pair has been eliminated out of the equation the imperative is to find another one and procreate. If a mate disappears, a hawk is not likely to wait until it comes back.

"I can only speculate on what happened this time. When the original female was injured and taken to the Raptor Trust, the second one obviously filled the niche pretty quickly. I'm assuming that's Domino's bird, the one-eyed bird from last year. Well, that bird could very well have been the dominant of these two females. So let's say that first bird came back to the park after I released her. Well, the second female probably evicted her. But once the second one was gone, the first one could come back. Obviously I have no idea where the first one has been all these years. I can only speculate again. Maybe she nested somewhere else and lost *her* mate. We'll never know.

"The bottom line in all this is procreation. They've got to keep the species going. In the natural world many individuals are sacrificed in that process, as you know. We're about the only species that puts so little value on the whole picture. We put it on individuals—that's what's so damned important for us. In the rest of Nature, individuals are superfluous. The species must endure."

SCENE THREE

Rapture Redux

Under the watchful gaze of the hawkwatchers Pale Male and First Love had a romantic rendezvous at a famous New York place for romance: the Hotel Carlyle. But while other lovers conduct their dalliance within, this pair chose the green copper tower for their lovemaking. No one had ever seen the hawks perch there before, though the handsome structure at Madison Avenue and 76th Street was plainly visible from the hawk bench.

By the beginning of April the sex romp was over. Time for sedate exchanges at the Fifth Avenue nest: one bird in, the other bird out. There were, to be sure, occasional flare-ups of passion, a quick show of affection on Woody's antenna or atop the Stovepipe. But the couple's subscription to True Romance had expired. Good Housekeeping was on its way.

On April 24th the hawk pair ceased sitting, and each took turns perching at the edge of the nest, making those up-and-down feeding movements the hawkwatchers understood clearly: the eggs must have hatched.

May 5th was a capital letter day in the Bird Register

AT LAST—BABY RED-TAILED HAWKS!! First baby sighted through Charles' telescope at 4:00 p.m. Jim

Lewis sees a second fuzzy white head a minute later. In attendance were Merrill, Marcia, Sharon, Mary and others. Lots of excitement.

Holly Holden and Jim Lewis

First Love had become Mom II. A new generation of red-tailed hawks was on its way.

SCENE FOUR

An End and a Beginning

Pale Male and his first love successfully raised three chicks in the Fifth Avenue nest the year they were reunited. Though scientific accounts state that the male red-tailed hawk provides most of the food for the female and nestlings, the faithful hawkwatchers agreed that the new hawk-mother brought in almost as many birds and rodents to the hungry young as her mate. By the same token, Mom II seemed to spend more time away from the kids than her predecessor had—the able huntress was sometimes absent for several hours at a time. "A modern career woman," someone quipped.

The following year two chicks hatched toward the end of April. They both fledged on June 12th, one at 5:58 a.m., the other at 7:36 p.m., and Frederic Lilien, the young Belgian documentary filmmaker, was there to document both flights. He had been filming the Fifth Avenue hawk family for three seasons now and was having trouble finding a conclusion for his film. He was as hooked on the hawks as any of the hawkwatchers, and, besides, something new and exciting kept happening. Then a real ending came along, though it was also a beginning.

It happened in mid-October, on a day when the Tupelo tree in the meadow beyond the Azalea Pond was full of migrating flickers and thrushes gorging on its ripe blue berries. Eleven species of southbound warblers were stopping over in Central Park that day, as well as two scarlet tanagers, an indigo bunting, a rose-breasted grosbeak, and still others, according to Tom Fiore's report in the Bird Register. Flocks of white-throated sparrows were already streaming into the park for their winter stay. That was when a mature female red-tailed hawk was sighted on a ledge near the roof of the Metropolitan Museum of Art at Fifth Avenue and 81st Street.

The hawk was wearing a U.S. Fish and Wildlife Service band on her right ankle. But there was no need to resort to telescope or binoculars to read the number. She was dead.

In the talons of the bird wearing band #1387-38569, the one we had once called First Love and then Mom II, lay a partly eaten pigeon. It was almost certainly the cause of the hawk's death, for nearby building managements were known to periodically set out poisoned pellets for the pigeons congregating in their vicinity. This time they had killed the best natural pigeon-exterminator they could have found.

"Sad times," read an unsigned entry in the Bird Register that week. "Bird Lovers unite!" wrote Rebekah Creshkoff. "Send letters to the mayor urging an immediate moratorium on poisoning pigeons in NYC."

The press picked up the story. "The most famous red-tailed hawks in the world," the *New York Post* declared, recounting, the history of Pale Male and First Love in heart-breaking detail. Mary Tyler Moore appeared on the early morning news and expressed her distress.

THE BODY OF THE HAWK was removed from the museum ledge on a Tuesday. Two days later a female red-tailed hawk was spotted on a window railing just south of the hawk building. The bird had a dark head and a very distinct belly-band.

A few moments later Pale Male lazily floated into view. He landed beside the larger, darker female and almost at once the birds took off together. They soared above Fifth Avenue in the manner of courting red-tailed hawks, seeming to hang in space for a few long seconds before sailing into Central Park. Over the model-boat pond they flew, almost directly above the heads of a group of hawkwatchers assembled there. The birds circled the pond and then landed in a large cottonwood-poplar just to its north. The new hawk's beak was grayer than her predecessor's and had a tinge of color when the sun hit it the right way. The hawkwatchers began to call her Blue.

Curtain Call

Even yet thou art to me
No bird: but an invisible thing
A voice, a mystery.

WILLIAM WORDSWORTH

The Earlybirds were about to celebrate their first birthday. Though it included some experienced birdwatchers by now, the weekly walking group remained faithful to its original mission: to introduce hawkwatchers to the joys of watching *other* birds and studying nature in Central Park. The group assembled bright and early on Wednesday mornings at the old hawk bench.

Jane Koryn was one of the original members. Hers was the living room window from which the numbers on Mom's band had been deciphered the year before; hers were the coffee and flaky Danish pastries that had sustained the hawkwatchers through the tense morning hours of waiting for the first batch of red-tailed hawk babies to fledge. She had remained a faithful hawkwatcher thereafter, and as the Red-tail Show–Round 2, starring Pale Male and First Love, played at the model-boat pond, she followed it with undiminished ardor. After a full year of weekly bird walks she was on her way to becoming a good birdwatcher.

Others on the Earlybird roster as it entered its second year included: two psychiatrists, a lawyer, a housewife, a retired teacher, a retired fund-raiser, a graduate student in public health, an executive of a small philanthropic foundation, and an aspiring novelist. Tom Fiore and other Regulars showed up now and then, and even an occasional Big Gun like Marty Sohmer.

Dorothy Poole, the group's leader, was a major draw. She has a great ear, a fine eye, contagious enthusiasm—a top-notch birdwatcher. The experienced birdwatchers knew they would see interesting birds on a walk with Dorothy: esoteric sparrows, finches, difficult warblers. The beginners and intermediates learned to look more closely, to listen more carefully, and, simply by Dorothy's example, to care more deeply about the fate of birds in a city park. Dorothy is not only a birdwatcher, but a bird-lover.

Most Wednesdays the Earlybirds head straight for the Ramble. On the morning of June 6th, however, they decided to start at Turtle Pond, a small body of water on the south side of the Great Lawn, just at the base of Belvedere Castle. A haven for a bunch of mallards, a few nesting red-winged blackbirds, one or two pairs of song sparrows, some dragonflies, and not much else, Turtle Pond does not usually attract birders.

But a few months earlier the little pond had been completely drained as part of the Central Park Conservancy's restoration of the Great Lawn that year. It was to be enlarged, deepened, and re-landscaped, with a special island to be created at the pond's western end for the turtles' basking pleasure. Work was scheduled to start in the middle of June.

The turtles—thirty-eight in all, mostly red-eared sliders and snapping turtles, but also a few musk turtles, painted turtles, and cooters—had been netted, checked for shell rot (a bacterial infection), weighed, measured, marked with coded notches on the edge plates of their shells, and then "temporarily relocated." That is, they were all dumped in the rowboat lake.

Now Turtle Pond was a pond no more. It was a mud flat with an occasional scummy puddle dotting its litter-strewn surface, an eyesore to most passersby. Not to birdwatchers, however. For shortly after the bottom of the pond was exposed, a whole slew of interesting birds had begun to show up—herons, egrets, and a variety of shorebirds—to feed on newly revealed riches there. Some of these birds, the glossy ibis, for example, and the least sandpiper, were species rarely seen in Central Park. It was at the drained Turtle Pond that new bird-watcher Ilenne Goldstein had seen her controversial little blue heron a few weeks earlier.

It was at the drained Turtle Pond that the Earlybirds saw a killdeer on the morning of June 6th.

Though an everyday bird in a great many places, the killdeer is uncommon in Central Park. It prefers bare, open ground for foraging and nesting, perhaps, as experts have suggested, because it is easier to see predators approaching in that sort of terrain. Bare ground, however, is precisely the sort of thing a park strives to eliminate. Luckily for the Wednesday morning Earlybirds, bare ground is exactly what a pair of killdeer found at the base of Belvedere Castle where Turtle Pond had once glistened.

THE KILLDEER IS a robin-sized member of the plover family with longish legs, a black bill, and two black bands elegantly draped over a snowy white breast. Though officially a shorebird, the killdeer is a bird of fields and meadows, pastures and dry uplands, and is often found in inland areas far from water. Needless to say the killdeer does not prey on antler-bearing mammals, as its name improbably implies; as with the chickadee, the pewee, and the whip-poor-will, this bird's name derives from its call: a "half-petulant, half-plaintive cry—Kill-deee, kill-dee!" as Frank Chapman, a noted ornithologist of the past, described it in 1912.

Charadrius vociferus, the killdeer's scientific nom de plume, is a common and abundant bird throughout most of the United States. According to Jerome Jackson, a biologist at Mississippi State University, the species is probably more abundant today than before the arrival of European man. That is because it thrives in the sort of disturbed habitat so common these days—construction sites, suburban subdivisions, abandoned lots, rubbish heaps, and the like.

The killdeer's particular specialty is the distraction display, that exaggerated, dramatic performance of feigned injury or exhaustion by which parent birds of certain species call attention to themselves in order to lead predators away from their nest or young. The killdeer's broken-wing act is masterful, "the epitome of pathos and verisimilitude," according to Christopher Leahy's *The Birdwatcher's Companion.*

But the broken-wing display is not the only trick in the killdeer's bag. The bird has a repertory of dramatic numbers, choosing one or another depending on the type of predator it encounters. If a big beast like a cow appears out of nowhere and is about to step on the killdeer's nest or stumble on its young, there's no time for Method acting. In this case it produces its "ungulate display." As the cow approaches, the parent bird positions itself directly in front of the animal, all the while flashing its wings and making a big racket. This startles the cow. It stops in its tracks and doesn't take that fatal step. In the ungulate display, the killdeer is not pretending to be sick or injured; as it flies and flashes its wings in the very face of a big and dangerous animal, it might be said to be feigning insanity!

It was one of the group's two experts on feigned insanity who first spotted the killdeer on the morning of June 6th. Though the bird was out in the open, its color and markings blended in so well with the cracks in the earth and the scattered rocks and debris all around that it was effectively camouflaged. The sharp-eyed psychiatrist had to resort to using items of litter as landmarks to point out the bird's location: "Look a few feet south of that old 78-rpm record sticking out of the mud, then find the midpoint of a line between the orange traffic cone and the rusty beer can to its right."

Soon everyone had located the bird, and they all agreed that there was something odd about the way it was sitting. When another killdeer came into their field of vision and proceeded to sit in the same spot, while the first took off, the idea occurred to everyone at the same moment: Could there be a killdeer nest in the middle of Turtle Pond?

A few minutes later, at the edge of the pond, the Earlybirds had their answer: The nest itself was in clear view. It was a mere depression in the ground measuring a few inches in diameter—a scrape, as it is called in bird books. There were a few pebbles and bits of debris messily scattered about it, and in its center reposed four oval-shaped, buff-colored eggs, irregularly spotted with brownish splotches.

The Earlybirds' pleasure at discovering the nest was short-lived, however. Any day now, they realized with horror, the huge machines already standing at the sidelines would begin excavating the bottom of Turtle Pond. And there at the western end, almost exactly at the spot where the turtle-basking island was to be created out of the excavated dirt and rocks, sat the killdeer nest.

A few hours later Ursula Hoskins, the project coordinator for the Turtle Pond restoration, received a call from one of her assistants. "We've got trouble," he told her breathlessly. On his way to the Castle early that morning he had run into a group of birdwatchers who showed him a nest in the middle of the pond. When they heard that construction work on Turtle Pond was scheduled to begin in a week or two they seemed to get excited. No, no, it would have to be delayed until the eggs had hatched and the babies were ready to fly, they insisted.

Ms. Hoskins knew all about the contretemps between the Conservancy and the Central Park birdwatchers back in the early eighties. It had involved a restoration project too, and had ended up with great numbers of people signing a bitter protest. This could start another Era of Bad Feeling. When one of the birdwatchers called her that afternoon, she promised she would do what she could to delay the construction work. How much time would they need?

She was sent a fact sheet with the pertinent information: killdeer incubate their eggs for twenty-four to twenty-six days. Like many shorebird species that nest on the ground, killdeer are precocial—the chicks are hatched with a fine cover of down, and can walk and feed immediately after hatching. But they cannot fly until twenty-five days after hatching. Since the pond area was now surrounded by a tall cyclone fence, the chicks would be trapped if construction began before they could fly. If the eggs hatched toward the end of June as the birdwatchers calculated, the chicks should be airborne by July 25th.

There was another compelling reason to treat the inconvenient nest with care: the Migratory Bird Treaty Act of 1918 was still the law of the land. The same law had helped the Fifth Avenue red-tails keep their nest intact. Now it might save the lives of a killdeer family.

"I'll do what I can," Ms. Hoskins said, undoubtedly praying that the inevitable delays accompanying every construction project would be even longer than usual.

THE EARLYBIRDS WROTE their finding in the Bird Register at the end of their walk. From that day on the park's birdwatching community monitored the nesting killdeer daily, while the Earlybirds checked them out every Wednesday morning. One week went by, two weeks, three weeks, and the machines continued to stand idle at the edge of the drained pond. The killdeer took turns sitting on the nest, and one could regularly be seen foraging near the remaining puddles of water. Though grass soon began to grow on the bare mud flats, obscuring the nest and the sitting bird completely, just about every day someone still managed to see one or another of the killdeer foraging nearby.

On June 29th, a Saturday, as both killdeer emerged for a moment from the tall grass quite near the spot where the nest had first been sighted, something small and downy, mostly whitish in color, skittered out for a moment and then quickly disappeared again. A chick! Before long two other chicks had been sighted, and that seemed to be the grand total—one egg may not have hatched, or one of the nestlings hadn't made it.

On the Fourth of July, Tom Fiore saw three young killdeer and one adult on the west side of Turtle Pond. He wrote in the Register: "The young no longer downy but still small (nearly ½ adult size) and probably unable to fly yet. One stretched its wings up and they were no more than 2 inches long . . . very cute."

The birdwatchers rejoiced. So far so good. Now, if only the little bird family could hang in there for another three weeks or so. Then let the digging begin.

On July 13th Tom Fiore wrote in the Bird Book: "3 young killdeer and at least 1 adult in Turtle Pond. The young now about ⅔ adult size and rapidly gaining the adult plumage, but apparently still unable to fly."

On July 17th when the Earlybirds arrived at the Castle to check the killdeer, a new sound could be heard breaking the morning quiet. The machine was big, yellow, noisy. In front was a cab on caterpillar treads, in back a huge boom with a toothed bucket—a backhoe. It had begun digging at the east end of the pond.

After some anxious searching, the birdwatchers found the killdeer family. The three chicks were feeding beside a puddle near the center of the former pond. They completely ignored the metal monster snuffling and snorting nearby. One of the parents was standing on a large rock nearby, looking down at the scene, supervising.

Another call to Ursula Hoskins: The babies need only a week or so before they can fly out of there. Can anything be done?

Ms. Hoskins was horrified. "We walked all through the area on Monday, looking for the killdeer nest. Four of us searched, but we found no sign of it. We really had to start excavation on Wednesday, but we decided to begin at the east end of the pond, just in case the birds were there somewhere. That would give them a bit more time."

THE GRASS WAS now high and thick on the pond bottom; even the best birdwatchers in Central Park had a hard time finding the killdeer. Nevertheless, now more than ever the Regulars persisted in their daily search for the parents and chicks. Everyone was worried. Each day the machines dredged out and carted away more and more of the former mud bottom of Turtle Pond. Steadily the killdeer family was confined to a smaller and smaller part of the total area.

When the Earlybirds arrived at Belvedere Castle the following Wednesday, the scene below stopped their breath. Two-thirds of the pond had been excavated and was now dry as dust. The backhoe was already at work, and it was moving slowly and inexorably westward.

It didn't take the birdwatchers long to locate the killdeer.

The three big chicks were feeding near the last remaining large puddle while just a few feet away from them the big yellow digging machine was thrusting and poking at the earth. The parent bird was standing on the same rock she'd been seen on the week before. But today her mood was different: She was having a fit.

I say "she" because everyone assumed the bird on the rock was the mother bird. In fact, male and female killdeer look exactly alike. The bird on the rock screaming down to the chicks to "Look out! Come right over here! Are you crazy or what?" could just as well have been the father. But whatever the sex of the frantic parent, one thing was clear: The young paid no attention whatsoever. They just continued feeding in the rich, gooey mud.

The parent bird's distress was painfully obvious. But the children's cheerful defiance, their delicious adolescent confidence, their sense of invulnerability in the face of danger touched a chord in everyone. One of the retired birders chuckled and said, "Will you look at those kids!"

It did seem a classic case of parental overanxiety. The young killdeer were fine. By the time the workers knocked off in midafternoon, both parents and young were peacefully foraging in the westernmost quarter of the pond—the last unexcavated section. At least it still included one large, murky puddle.

Two days later Charles Kennedy gained a new understanding of the realities of killdeer parenthood, though it wasn't the killdeer that drew him to Turtle Pond that day. It was a juvenile red-tailed hawk. The second year's batch of Fifth Avenue red-tail babies had fledged in mid-June, with even bigger crowds in attendance than the year before. The three young hawks had spent the following month learning how to be successful killers, again before a rapt audience of hawkwatchers and bystanders.

Now one of the three fledglings was standing on the dry, excavated ground smack in the middle of Turtle Pond, demonstrating how well he had learned. He'd managed to catch a small rat, and Charles was trying to photograph the event. He had been following the clumsy and inexperienced red-tail from place to place, watching it pounce . . . and miss, again and again. Finally, at Turtle Pond, the young hawk had gotten his act together.

Charles had arrived at about 5:00 p.m.—the construction workers were long gone after digging up the last big puddle that day. All that was left was a little grassy area at the very western end of the pond.

At about 5:30 it began to rain lightly. Just as the hawk finished his meager meal, the drama happened. In Charles' words:

> The young hawk still looked a bit hungry—the baby rat had just whetted his appetite. His crop was almost flat. He flopped around and then flew toward the west end of the pond, where he landed on the ground again. Suddenly I saw Mama Killdeer pop up out of nowhere just a few feet in front of the baby hawk. My heart sank—I had thought the killdeer were long out of there.
>
> Then, amazingly, she charged! Right up to him, bill to beak. And she proceeded to carry on, yelling and yammering and screaming at him. And suddenly she did something new: Zing! She began to flash her wings, flash, flash, right in the hawk's face. You saw this flag—the killdeer's unbelievably bright rufous and white underwing flashing open.
>
> The baby hawk looked startled. He made an ineffective little pass at her, which she fielded gracefully, like a bullfighter holding out a cape at a poor, dumb bull. Then she flew off. He

flew after her, but she easily escaped. Obviously she had led him away from the babies. I've never seen such bravery!

There in the middle of New York City, miles from the nearest country pasture, Mama (or Papa) killdeer had treated the red-tail on the ground like a blundering bovine—the classic ungulate display. And it worked like a charm. Where, meanwhile, were those fearless and confident teenagers? Cowering behind a clump of grass, no doubt, while Papa (or Mama) made sure that for all their brash self-assurance they didn't end up as a hungry hawk's dinner.

TWO DAYS LATER the killdeer family left Turtle Pond for good. On Saturday, August 3rd, as Tom Fiore was birdwatching in the Conservatory Garden at the north end of the park, who should he run into behind the leaf-mulch compost heap but the Turtle Pond killdeer, all five of them. They were combing the rich earth for worms, ants, grasshoppers, caterpillars, spiders, bugs, beetles, and other invertebrates before taking off for good, God knows where. The red-tail fledglings hung around a few more weeks and then they, too, were gone, God knows where. Only unhappy stories have real endings, after all. You never find out how the good stories end.

Afterword

When I first started writing about the Fifth Avenue hawks, I had no way of knowing that the story I'd stumbled on was barely beginning. As the years went by, the hawk drama took so many unexpected turns that I had to start carrying a little crib sheet to keep track of who was who and what happened when. (For readers who want to keep things straight, there's some help in the index under the heading Pale Male).

The book had already gone to press when its saddest scene was played out: my romantic heroine, First Love, the big beautiful female hawk who'd hooked up with Pale Male when he was just a wild teenager, and whose reunion with Pale Male after years of separation formed the dramatic climax of the story, was found dead on a ledge of the Metropolitan Museum of Art one late October day, the remains of a poisoned pigeon still in her talons.

I barely had time to fax a new and darker ending to the printer. I softened the blow (for myself as much as for my future readers) by reporting the appearance of a new female red-tail at Pale Male's side two days after his mate's death. We called her Blue. The birdwatchers marveled as the new couple performed courtship maneuvers over the Model-boat

Pond. A quick operator, that Pale Male, we agreed. "His guyness," Charles Kennedy began to call him.

Pale Male and Blue were sitting on eggs by the publication date of *Red-Tails In Love* the following March. By the middle of April, for the fourth year in a row there were fuzzy white chicks popping up and down in the nest high above Fifth Avenue.

New faces dominated the hawk bench. The old-timers who'd been there when the story began—Charles, Tom, Norma, Anne, Jim, Holly and Blanche, Jane of the early-morning pastries—still dropped by regularly. But the torch had been passed.

As Pale Male and Blue fed their chicks high over Fifth Avenue, the passionate observers who spent hours on hawk-watch duty day after day, the ones who arrived before sunrise and stayed until dark were now Annie and Ben (who discovered a red-headed woodpecker roost hole a few months later), John and Patti (who joined the Wednesday Earlybirds), Noreen (who kept track of the sunset fly-outs and sunrise fly-ins of a pair of wintering long-eared owls), Lee, Paul, Stephanie, Denise, Tommy, Tom, John, Jeanine and Mike, Janet, Susan, Ron, and baby Griffin in a backpack.

The Bird Register gained a new correspondent: Michael, a latter-day Thucydides, who filled its pages with dramatic histories of the Wars between the Red-tails and the Crows. Merrill Higgins continued to set up his telescope for the world at large. He'd upgraded to a Nikon Fieldscope now, top of the line. Frederic Lilien was also a regular presence at the hawk bench, still working on a film about the Fifth Avenue hawks. Some of his spectacular footage had begun to appear in various nature documentaries.

That year the ever-suspenseful fledging drama departed from the usual script. The first chick sailed off in a high wind on the afternoon of June 2nd, Fledger #2 took his first flight very early the next morning, but the third chick—a male, by the look of him—stayed put. Though hawk nestlings usually fledge at intervals of one or two days, this one didn't leave the next day, or the next or even the next.

The fledglings of years past had never returned to the nest once they'd taken flight. But one week after the day that our second hawklet fledged, I looked up and saw a surprising sight: three babies in the nest instead of one. The two previous fledgers had come back to visit their sibling. Was it to offer moral support? Maybe they just came for dinner. When Pale Male arrived with a pigeon an hour or two later, all three tucked in.

By now the mood at the hawk bench had turned from light-hearted anticipation to doom prognostication: Something was terribly wrong. It took ten days for the last nestling to get off his duff and sail into the park. Since he chose a driving rainstorm for his lift-off, most of the regular hawkbenchers missed it. Only Stephanie Schmidt, a new and very determined hawk addict, was there at 12:31 p.m. to see the dilatory hawklet take to the air.

The young woman was soaked to the skin but ecstatic as she watched the fledgling land awkwardly in the crown of a tall tree near Fifth Avenue and 76th Street. "For about 5 minutes (or maybe 10) he flapped, seemingly trying to retain his balance while also being 'attacked' by a blue jay," she wrote in the Bird Register. Evidently she remained beside the tree for at least two-and-a-half hours herself, singing in the rain, no doubt, watching over the baby hawk.

(When asked about the long interval between the second and third chick's maiden voyages, hawk expert Charles Preston responded: "It's unusual but it happens. It's possible that the first two left prematurely because of the wind.")

My book had just arrived in bookstores that spring and a number of readers made their way to the Boathouse to look for the Bird Register. Some wrote entries, among them Ellen Metzger from Michigan who reported seeing one of the immature red-tails perched atop an Ellsworth Kelly sculpture in the roof garden of the Metropolitan Museum. I met a few other readers wandering through the Ramble. How did I recognize them? They were using the book's endpaper map as a guide. There's no denying that running into people with one's book in hand is a pick-me-up of the highest order.

Readers and park passersby asked one question again and again: What becomes of the fledglings at the end of the season? We know that as the next breeding season approaches, adult hawks begin to defend their territory from all birds of prey, even their own offspring. Where then, do the juveniles go? It will remain a mystery unless the chicks are banded, but we do have a clue.

Red-tailed hawks rarely breed until their second year. That's when their brown tails turn red. (Pale Male, a browntail when he first found love, was an exception to this rule.) Two years after that glorious April day when chicks were first sighted in the Fifth Avenue nest and epic madness reigned at the Model-boat Pond, just when those first chicks might have been old enough to build a nest of their own, the following entry appeared in the Bird Register:

> A red-tailed hawk [adult] was seen breaking off twigs and bringing them to a ledge above the left uppermost window of Mt. Sinai Medical Center at Fifth Avenue and 100th Street. Stay tuned for further developments.
>
> Tom Fiore

Ten days later, another birdwatcher reported:

> Much nest-building activity by Mt. Sinai Red-tails. One bears an uncanny resemblance to Pale Male.
>
> S. Hammer

Nothing came of the Sinai nesting attempt, neither that year nor the next when the pair tried again. But it does not seem unlikely that one of the hawks building a nest there had been a former Fifth Avenue fledgling trying to recreate his first home, a chip off the old block in behavior as well as appearance. Unfortunately the Mt. Sinai site lacked a crucial element present at 74th and Fifth: the anti-pigeon spikes to hold the nest firmly in place. Without them the accumulation of sticks eventually blew away.

Even as the hawk story was creating new suspense and drama, the park's human community generated some heart-stopping news of its own. On March 23, 1998, the same week this book was officially published, one of its central figures, Tom Fiore, found himself in terrible trouble. Four days earlier he had gone on a short birdwatching trip to Colombia with three friends, one of them, Pete Shen, a Central Park birder as well. While searching for a rare bird in the Andean foothills, the group was seized by a band of armed anti-government guerrillas and taken as hostages. (Colombia is sometimes called "The kidnap capital of the world".) A period of almost unbearable anxiety began for the Central Park birdwatching community.

The Americans' binoculars and the tape recorder they carried for capturing bird songs aroused the rebels' suspicions. Only spies carry such equipment, they declared, unfamiliar with the very concept of birdwatching. Their commander appeared on Colombian television five days after the abduction and threatened the Americans with execution if they turned out to be undercover U.S. intelligence agents.

The Central Park birdwatchers took swift action, securing affidavits from the Linnaean and Audubon Societies confirming that Tom and Pete were members in good standing. (Friends in other cities rounded up similar documents for Todd Mark and Louise Augustine, the other two hapless birders on the trip.) A packet of these and other birdwatching bona fides, including pages from the Bird Register with Tom's lengthy bird lists, were sent to Bogota by a special courier to be forwarded to the rebel camp. A copy of *Red-tails in Love* was included in the collection, with Tom Fiore's contribution to the appendix—"Birds Through the Year in Central Park"—marked for special attention.

The birdwatchers' evidence-gathering efforts may have helped: several days after the materials were received in Colombia the death threat was lifted. But the Central Park anxiety level remained high. Reports indicated that hostages of Colombian terrorists were often held for months, even years, before they were released.

Nine days after his capture Tom Fiore pulled off an extraordinary coup: he escaped from his armed guards, making his way to freedom through the jungle "with the wherewithal that might do Indiana Jones proud," as a *New York Times* reporter wrote the following day.

On April 4 an entry appeared in the Bird Register that began to restore normality to the Central Park birdwatching community:

> It's great to be back in Central Park again today—showing the red-headed woodpecker to reporters for Colombian TV. We hope for the early and safe return of the 3 other birdwatchers still being held hostage in the mountains of Colombia. Thanks to all . . .
>
> Tom Fiore

The others were released three weeks later, and a nightmarish episode came to a happy ending.

As for the other Regulars: Four years after Sharon Freedman began her City Hawk Watch at Belvedere Castle, it was taken over by the Urban Park Rangers as an official park activity. The following year Sharon moved to North Carolina.

That same year Charles Kennedy switched his allegiance from the 88-pace brook near Balcony Bridge to a new butterfly garden north of the Boathouse. Charles and a few other Regulars had helped clear an overgrown little wilderness there, and with the help of the Conservancy's horticulture expert Regina Alvarez they planted it with lepidoptera-friendly flowers and shrubs. Butterflies showed up in gratifying numbers and so did hosts of other insects. On eight evenings in August and September, with the help of a night light set up at the garden's northeast corner, a group of Regulars identified 33 species of moths, including The Penitent, The Grateful Midget, and Pale Beauty.

Charles became the garden's unofficial caretaker, so steady a presence there that it became generally known as the Kennedy Garden. It wasn't long before it became a hangout for the park's nature congregation.

Butterflies had long attracted the interest of Central Park's nature lovers. That year Nick Wagerik inspired the same set of worshipers to focus their attention on an additional insect order, the Odonata—dragonflies and damselflies.

Turtle Pond had been enlarged and re-landscaped as part of the Great Lawn restoration and its Odonate population was burgeoning. Many of the faithful band gravitated there regularly to study dragonflies with Nick, among them Norma Collin, Mo and Sylvia Cohen, Chris and Marianne Girards, Janet Wooten, along with talented newcomers like Ed Lam, a young artist whose work appears regularly in *The New York Times,* and Davie Rolnick, an eager seven-year-old with a particular passion for bugs. By the end of the season many had acquired the arcane skills necessary to distinguish an ebony jewelwing from a slender spreadwing, for example, or a blue dasher from a spot-winged glider.

Even casual strollers stopping by out of curiosity learned to distinguish a dragonfly from a damselfly: The first is a large, stout-bodied insect that holds its wings out like an airplane when at rest. The latter is a closely related but more diminutive creature that generally perches with its wings folded up against its abdomen or slightly above it. (A list of Central Park's Odonata, newly compiled by Nick Wagerik for this edition, appears in the Wildlife Almanac at the end of the book.)

IT'S WINTER NOW. The dragonflies are gone. The earliest spring migrants, flocks of noisy grackles, have not arrived. Of the year-round residents neither the titmouse nor the cardinal has uttered its spring song yet, at least not according to the Bird Register. Yet Pale Male and Blue have already begun bringing twigs to the Fifth Avenue nest, refurbishing, redecorating. Soon they'll be mating on top of Woody's antenna, or on the black smokestack, or the tower of The Carlyle. If all goes well there should be chicks in the nest by the time this edition is out in the world. The park will be full of warblers, vireos and thrushes by then, as well as birdwatchers checking them out. It's a grand spectacle. I can only repeat the invitation I made at the beginning of this book: Come to Central Park and see for yourself.

December 1998

CENTRAL PARK

A Wildlife Almanac

Birds Through the Year in Central Park

Tom Fiore

This list is based on sightings since 1990 and includes the 190 species found annually or nearly so. (Nearly ninety additional species have been recorded in the past century, some only once, others a few times in a decade, some not seen in many decades.) The most common or easily seen birds are shown in **bold** type. Those shown in standard type are fairly common in the proper habitat and season. Birds seen less easily or infrequently are shown in *italics*. All species known to nest are noted with an asterisk; uncommon or occasional nesters are noted with a cross. Emphasis in the calendar is on spring arrivals. Fall migration is shown in a more general way.

Birds of prey are shown in their own list, compiled by Sharon Freedman, who has devoted thousands of hours to studying the movements of raptors flying over the park.

A number of species may be found year-round:

Mute Swan†
Mallard*
Red-tailed Hawk†
American Kestrel
Ring-necked Pheasant†
Ring-billed Gull
Herring Gull
Great Black-backed Gull
Rock Dove (Pigeon)*
Mourning Dove*
Red-bellied Woodpecker*
Downy Woodpecker*
Blue Jay*
American Crow*
Black-capped Chickadee†
Tufted Titmouse†
Northern Mockingbird*
European Starling*
Northern Cardinal*
Song Sparrow*
House Finch*
House Sparrow*

JANUARY Up to fifty species, including our residents, may be present. Among these are regular winter-long visitors as well as others seen only occasionally; some will be found in fall.

FEBRUARY As longer days combine with any milder weather, some of our earliest migrants begin to arrive: **Red-winged Blackbird*** and **Common Grackle,*** both often quite vocal, and the secretive American Woodcock, most active near dawn and dusk.

MARCH Expected before winter's end, even in fickle weather:
Canada Goose Mainly flyovers; some may drop in for a rest.
Killdeer Sometimes numerous, yet may pass almost unnoticed.
Eastern Phoebe The first invariably comes by mid-month.
American Robin Numbers build to hundreds by month's end.
Eastern Bluebird and Eastern Meadowlark. Either is a prized find in the park!
Fox Sparrow A few may have wintered; often singing now.

Migration picks up noticeably around the vernal equinox:

Great Egret
Black-crowned Night Heron
Hooded Merganser
Belted Kingfisher
Northern Flicker*
Tree Swallow
Fish Crow (identified by voice)
Winter Wren
Golden-crowned Kinglet
Pine Warbler (often 1st warbler)
Louisiana Waterthrush (by month's end)
Dark-eyed Junco

APRIL The pace of arrivals quickens with each warm day; a few species will likely wander in unexpectedly early.

1st–10th

Red-throated Loon } mainly flyovers
Common Loon } mainly flyovers
Pied-billed Grebe
Great Blue Heron
Double-crested Cormorant
Brown Creeper
Marsh Wren
Ruby-crowned Kinglet
Hermit Thrush (often numerous)
Palm Warbler

Blue-winged Teal
Common Snipe
Laughing Gull
Yellow-bellied Sapsucker
Northern Rough-winged Swallow
Eastern (Rufous-sided) Towhee
Swamp Sparrow
Brown-headed Cowbird
Rusty Blackbird
American Goldfinch

11th–20th

American Bittern
Green Heron†
Chimney Swift
House Wren*
Brown Thrasher†
White-eyed Vireo
Blue-headed (Solitary) Vireo
Yellow-rumped (Myrtle) Warbler
Black-and-white Warbler
Hooded Warbler
Indigo Bunting
Chipping Sparrow
Field Sparrow
Savannah Sparrow

21st–30th

Snowy Egret
Solitary Sandpiper
Spotted Sandpiper
Least Flycatcher
Great Crested Flycatcher
Eastern Kingbird*
Barn Swallow
Bank Swallow
Cliff Swallow
Purple Martin
Veery
Wood Thrush†
Warbling Vireo*
Yellow-throated Vireo†
Blue-winged Warbler
Nashville Warbler
Northern Parula
Yellow Warbler
Black-throated Blue Warbler
Black-throated Green Warbler
Yellow-throated Warbler
Prairie Warbler
Cerulean Warbler
Prothonotary Warbler
Worm-eating Warbler
Ovenbird
Northern Waterthrush
Kentucky Warbler
Common Yellowthroat
Evening Grosbeak

MAY The greatest diversity of species now occurs, with a count of over one hundred possible on a good day! Bird song is now at its peak as well.

1st–10th

Black-billed Cuckoo
Yellow-billed Cuckoo
Whip-poor-will
Ruby-throated Hummingbird
Eastern Wood-Pewee
Alder Flycatcher
Willow Flycatcher
Red-breasted Nuthatch
Swainson's Thrush
Bicknell's Thrush
Gray Catbird*
Red-eyed Vireo†
Golden-winged Warbler
Lawrence's Warbler
Brewster's Warbler
Tennessee Warbler
Orange-crowned Warbler
Chestnut-sided Warbler
Magnolia Warbler
Cape May Warbler
Blackburnian Warbler
Bay-breasted Warbler
Blackpoll Warbler
American Redstart
Wilson's Warbler
Canada Warbler
Yellow-breasted Chat
Summer Tanager
Scarlet Tanager
Rose-breasted Grosbeak
Blue Grosbeak
Lincoln's Sparrow
White-crowned Sparrow
Bobolink
Orchard Oriole
Baltimore Oriole*

11th–20th

Chuck-will's-widow
Olive-sided Flycatcher
Acadian Flycatcher
Yellow-bellied Flycatcher
Gray-cheeked Thrush
Philadelphia Vireo (mostly fall)
Cedar Waxwing†
Mourning Warbler (often the very last species to arrive)

21st–31st

All the expected migrants have arrived. The earlier ones have mostly moved on; birds that will nest here are courting; many species have already set up housekeeping.

JUNE Migrants may continue to be seen even as summer begins; nesting birds and summer visitors are most likely now, however. The regular visitors: **Double-crested Cormorant, Great** and Snowy Egret, **Black-crowned Night Heron,** *Laughing Gull,* Chimney Swift.

JULY Birds sing less often now; many are busy watching over and feeding young. Some, notably robins, may attempt a second or third brood. After mid-month, some of the earliest "fall" migrants appear: Barn Swallow, Yellow Warbler, Louisiana Waterthrush, usually several other species by the end of the month.

AUGUST The southbound migration, more protracted than that of spring, gathers momentum. Some of our nesters, such as Eastern Kingbird, may depart. A Wood Duck or *Green-winged Teal* may arrive. Many birds become less colorful as adults molt into winter plumage and the more drab young birds are increasingly seen.

SEPTEMBER Exodus from the north is in full swing, with the peak diversity near mid-month. Most of the insect-eaters—swifts, swallows, vireos, warblers, and tanagers, as well as many of the thrushes—pass through. *Connecticut Warbler,* very rare in spring, is occasionally seen walking through the undergrowth this month.

OCTOBER Each cold-weather front pushes the remaining insect-eating species south; among the hardiest of these are: **Eastern Phoebe,** both kinglet species, Winter Wren, and Blue-headed (Solitary) Vireo. Numbers of sapsuckers, flickers, catbirds, thrushes, grosbeaks, buntings, and orioles swell and then diminish through the month. Large numbers of sparrows and other species arrive; some are regular winter-long visitors, such as **White-throated Sparrow** (often the most abundant winter songbird in the park, frequenting wooded areas and remaining into April), while others are occasional by winter, such as **Dark-eyed Junco.** Uncommon arrivals may include *Red-headed* or *Hairy Woodpecker;* either may select an area of the park in which to spend the winter. Snow Goose, seen mainly in high-flying skeins, often late in the day when northerly winds blow, can be spectacular—single-day observations of thousands have been made from the park. Rarities include *Clay-colored* and *Vesper Sparrow,* sometimes found among flocks of Chipping Sparrows.

NOVEMBER Migration continues with lower diversity, yet overall numbers of birds remain high. Some resident species show sizable increases: **Mallard,** the gulls, **Blue Jay,** chickadee, titmouse, and Song Sparrow. Migrants that may linger into winter include:

Great Blue Heron	Hermit Thrush
Black-crowned Night Heron	**American Robin**
American Coot	**Gray Catbird**
Yellow-bellied Sapsucker	Brown Thrasher
Northern Flicker	Cedar Waxwing
Red-breasted Nuthatch	Eastern (Rufous-sided) Towhee
White-breasted Nuthatch	Fox Sparrow
Brown Creeper	Swamp Sparrow
Carolina Wren	**Red-winged Blackbird**
Winter Wren	**Common Grackle**
Golden-crowned Kinglet	Brown-headed Cowbird
Ruby-crowned Kinglet	American Goldfinch

A rarely noted, yet possibly annual late migrant: *Horned Lark*. Two more-or-less annual visitors that may be discovered at any time from now into April: *Long-eared Owl* and *Saw-whet Owl*. More rarely, *Barn, Barred,* or *Great Horned Owls* are also seen. On the Lake, ponds, and especially the Reservoir may be found a variety of ducks; those seen regularly are:

Wood Duck	Canvasback } both have decreased
American Black Duck	Lesser Scaup } both have decreased
Northern Shoveler	**Bufflehead**
Gadwall (increasing)	**Ruddy Duck** (increasing)

More rarely seen: *American Wigeon, Northern Pintail, Greater Scaup, Ring-necked Duck, Redhead, Red-breasted,* and *Hooded Mergansers.*

DECEMBER Migration is for the most part over, yet each cold wave may bring a few more migrants or visitors. In some years, *American Tree Sparrow, Common Redpoll,* and *Pine Siskin* appear. The season is highlighted by the annual Christmas Bird Count, during which participants covering the park may turn up a late lingering migrant or a roosting owl. Some bird-watchers go into hibernation now while more hardy folk gather at the bird-feeding stations, with the occasional foray to the Reservoir in search of a rare duck or gull.

Butterflies of Central Park

Nicholas Wagerik

This is a summary of my butterfly observations in Central Park. The names used for butterflies are the official names of the North American Butterfly Association. I have seen fifty-three species in the park.

Abbreviations for frequently cited locations:

CG	Conservatory Garden
EM	East Meadow
NW	North Woods
R	The Ramble
SF	Strawberry Fields
SG	Shakespeare Garden
TP	Turtle Pond
WM	Wildflower Meadow south of Loch

PIPEVINE SWALLOWTAIL *(Battus philenor)*

June 8, 1987, one, R; July 17, 1987, one, R; July 26, 1989, one, CG; July 30, 1989, one, CG. This individual was seen on a plant called birthwort *(Aristolochia clematitis)*, a European plant in the same genus as the pipevine. On August 24, 1989, I observed a large caterpillar of the pipevine swallowtail on this plant. September 1, 1995, one, CG; October 11, 1995, one, WM; August 5, 1996, one, CG (August 11, 1996, caterpillars and eggs on birthwort, CG).

BLACK SWALLOWTAIL *(Papilio polyxenes)*

Uncommon. Earliest: April 23, 1995. Latest: September 29, 1988. Most often seen in CG. Also seen SG, SF.

EASTERN TIGER SWALLOWTAIL *(Papilio glaucus)*

Fairly common. Earliest: May 22, 1994. Latest: August 31, 1992. Maximum: July 21, 1994, six. Seen throughout the park, most easily in CG.

SPICEBUSH SWALLOWTAIL *(Papilio troilus)*

Much less frequently seen than Tiger Swallowtail. Earliest: May 17, 1993. Latest: September 9, 1996, four. Seen most often in R and CG. Also seen in NW and SG.

CHECKERED WHITE *(Pontia protodice)*

September 3, 1996, one, SG.

CABBAGE WHITE *(Pieris rapae)*

Very common. Earliest: April 4, 1997. Latest: November 24, 1993. Maximum: July 17, 1987, 135.

CLOUDED SULPHUR *(Colias philodice)*

Less often seen than the Orange Sulphur. Earliest: April 22, 1990. Latest: November 14, 1994.

ORANGE SULPHUR *(Colias eurytheme)*

Fairly common. Earliest: April 18, 1995. Latest: November 22, 1994.

CLOUDLESS SULPHUR *(Phoebis sennae)*

Summer of 1987 (exact date lost), one, SF; September 7, 1989, one, EM; September 6, 1995, two: one at West Drive at 90th St. and one at CG; September 10, 1995, one, Pinetum; October 6, 1995, one, West Drive at 96th St.

HARVESTER *(Feniseca tarquinius)*

June 26, June 28, June 30, 1995, one, NW.

BANDED HAIRSTREAK *(Satyrium calanus)*

Common in NW. Also seen in R and once in CG. Earliest: June 17, 1994. Latest: July 18, 1995. Maximum: July 1, 1994, twenty.

WHITE M HAIRSTREAK *(Parrhasius m-album)*

August 31, 1992, one, SG; June 2, 1994, one, West Drive near 62nd St.

RED-BANDED HAIRSTREAK. *(Calycopis cecrops)*

Usually uncommon. Earliest: May 23, 1995. Latest: September 24, 1992, five. Seen in CG, SF, SG, R.

GRAY HAIRSTREAK *(Strymon melinus)*

Rare, but regularly seen. Earliest: June 27, 1991. Latest: October 2, 1994. Most often seen in CG.

EASTERN TAILED BLUE *(Everes comyntas)*

Uncommon. Earliest: June 27, 1996. Latest: October 10, 1995. Seen in CG and on white clover in lawns.

SPRING AZURE *(Celastrina ladon)*

I have seen the spring brood of this species only once in Central Park, April 25, 1995, in the R. This was an individual of the form violacea. The summer brood may be a different species. It is common in Central Park. Earliest: May 28, 1994. Latest: September 14, 1995. Also seen on October 24, 25, and 28, 1994, in the CG. The October sightings, much later than the normal flight period, were of a freshly emerged individual, probably related to the warm fall weather.

AMERICAN SNOUT *(Libytheana carinenta)*

Irregular. Not found in some years. Earliest: May 25, 1991. Latest: August 21, 1989. Most often seen on or near hackberry trees in R, SF, and near Belvedere Castle. Often nectars on sweet pepperbush flowers.

VARIEGATED FRITILLARY *(Euptoieta claudia)*

June 13, 1991, one, CG; July 14, 1995, one, CG.

GREAT SPANGLED FRITILLARY *(Speyeria cybele)*

July 15, 1993, one, R; August 23, 1994, one, CG; August 27, 1994, two, CG; August 29, 1994, one, CG; September 20, 1994, one, CG; June 28, 1995, one, WM; July 4, 1995, one, WM; June 27, 1996, one, WM.

PEARL CRESCENT *(Phyciodes tharos)*

Uncommon. Earliest: May 12, 1993. Latest: October 25, 1994.

QUESTION MARK *(Polygonia interrogationis)*

Usually common. Earliest: April 9, 1991. Latest: November 15, 1993. Seen throughout park.

EASTERN COMMA *(Polygonia comma)*

This species varies considerably in abundance in the park and may be absent in a particular year. It is usually much less common than the similar Question Mark. In 1994 it was seen in much higher numbers than usual. Earliest: March 23, 1994. Latest: October 29, 1993. Most common in the R and NW but may be seen in many areas of the park.

MOURNING CLOAK *(Nymphalis antiopa)*

Fairly common. Most easily seen in spring and early summer in R and NW. Earliest: February 23, 1992. Latest: October 13, 1994.

COMPTON TORTOISESHELL *(Nymphalis vau-album)*

September 6, 1989, one, CG; September 3, 1995, one, CG; September 11, 1995, one, CG; September 14, 1995, one, CG (not the same individual as seen on September 11); September 19, 1996, one, TP.

RED ADMIRAL *(Vanessa atalanta)*

Earliest: April 23, 1994 and 1995. Latest: November 3, 1994. This species is often common and may be seen throughout the park.

AMERICAN LADY *(Vanessa virginiensis)*

Fairly common. Earliest: April 20, 1995. Latest: October 30, 1990. Maximum: August 2, 1991, seven.

PAINTED LADY *(Vanessa cardui)*

Earliest: April 22, 1990. Latest: October 9, 1993. This species varies greatly in abundance from year to year. Usually rare to uncommon. In 1991 this species greatly outnumbered the American Lady.

COMMON BUCKEYE *(Junonia coenia)*

July 31, 1990, two, CG; August, 13, 1994, one, CG; September 21, 1994, one, CG.

RED-SPOTTED PURPLE *(Limenitis arthemis)*

July 20, 21, 26, 1994, two; August 3, 1994, three: August 8, 1994. August 9, 1994. Seen in the R or nearby.

HACKBERRY EMPEROR *(Asterocampa celtis)*

July 20, 21, 1994, two, R.

MONARCH *(Danaus plexippus)*

Common to very common. Earliest: May 15, 1997. Latest: November 5, 1990. Maximum: October 22, 1994, 140, mostly nectaring on chrysanthemum flowers in CG.

SILVER-SPOTTED SKIPPER *(Epargyreus clarus)*

Common. Earliest: May 15, 1993. Latest: September 28, 1989. Most easily seen at the CG.

LONG-TAILED SKIPPER *(Urbanus proteus)*

August 20, 1994, CG; September 10, 1995, one, CG; September 14, 1995, one, CG (not the same individual as seen on September 10).

JUVENAL'S DUSKYWING *(Erynnis juvenalis)*

May 19, 1989; May 23, 1992; May 12, 26, 27, 1993; May 22, 1994; April 26, 1995. Seen in SG and NW.

HORACE'S DUSKYWING *(Erynnis horatius)*

August 1, 4, 1995, one, CG.

WILD INDIGO DUSKYWING *(Erynnis baptisiae)*

Uncommon. Earliest: May 26, 1993. Latest: September 15, 1995. Mostly seen in CG. I have seen this species ovipositing on blue false indigo *(Baptisia australis)* on two occasions.

COMMON CHECKERED SKIPPER *(Pyrgus communis)*

September 3, 1994, CG.

COMMON SOOTYWING *(Pholisora catullus)*

Uncommon. Earliest: May 27, 1990. Latest: September 28, 1994. Maximum: August 9, 12, 1994, four. Most often seen in CG.

SWARTHY SKIPPER *(Nastra lherminier)*

June 18, 1995, one, CG.

CLOUDED SKIPPER *(Lerema accius)*

October 1, 1995, one, CG.

LEAST SKIPPER *(Ancyloxypha numitor)*

I observed this species in 1994 from July 11 through September 28. In 1995 seen on June 20, August 1, and September 11, all in CG.

EUROPEAN SKIPPER *(Thymelicus lineola)*

June 20, 1995, one, WM; June 28, 1995, one, WM.

FIERY SKIPPER *(Hylephila phyleus)*

September 5, 1989, CG; September 20, 1994, CG; October 13, 1994, near Reservoir. August 21, 1995, one, CG. August 25, 1995, one, CG; September 15, 1995, two, CG; September 29, 1995, one, CG; August 24, 1996, one, CG.

PECK'S SKIPPER *(Polites peckius)*

Uncommon. Earliest: June 2, 1993. Latest: August 28, 1990. Most often seen in CG.

TAWNY-EDGED SKIPPER *(Polites themistocles)*

August 8, 1994, CG.

NORTHERN BROKEN DASH *(Wallengrenia egeremet)*

June 21, 25, 1991, two; June 27, 1991, July 5, 1993, September 15, 1995, one. All of these were seen in CG; July 5, 1996, one, WM.

LITTLE GLASSYWING *(Pompeius verna)*

June 21, 1991, CG; July 7, 1993, CG; July 14, 1995, WM.

SACHEM *(Atalopedes campestris)*

September 15, 1993; August 31, September 9, 10, 11, 14, 15, 1995. All were seen in CG.

HOBOMOK SKIPPER *(Poanes hobomok)*

June 3, 1991, CG, and a different individual on June 5, 1991, CG; June 2, 1993, CG; June 17, 1994, NW.

ZABULON SKIPPER *(Poanes zabulon)*

This species has two distinct broods. Earliest first brood: May 20, 1995. Latest first brood: July 8, 1995. Earliest second brood: July 16, 1991.

Latest second brood: September 21, 1994. The first brood is uncommon in the park and is sometimes absent. The second brood is common. Most often seen in CG.

BROAD-WINGED SKIPPER *(Poanes viator)*

July 22, 1989, two; July 25, 1989, two; July 26, 1989, two; July 29, 1989; July 25, 1994; July 26, 1994; August 9, 1994; July 24, 1995, two; August 1, 1995. Mostly seen on south shore of Belvedere Lake on sweet pepperbush flowers. Seen once in CG.

DUN SKIPPER *(Euphyes vestris)*

August 1, 1989, CG; September 23, 1992, SG; July 5, 1993, CG; June 10, 1994, R; July 21, 1994, CG; August 23, 1994, CG; August 29, 1994, CG; July 14, 1995, WM.

OCOLA SKIPPER *(Panoquina ocola)*

September 6, 1993, and a different individual on September 15, 1993; September 6, 1995, two; September 14, 15, 1995. All of these were seen in CG. September 2, 1996, one.

Damselflies and Dragonflies of Central Park

Nicholas Wagerik

Turtle Pond, the small body of water at the base of Belvedere Castle, was always Central Park's best location for observing damselflies and dragonflies. In the spring of 1998, after the pond had been enlarged and improved as part of the Great Lawn restoration, a greater number and variety of damsel- and dragonflies began showing up there. So did a growing number of regular observers. The following list of 50 species includes two that were first sighted in the park by some of these new enthusiasts. The common names used here were officially adopted by the Dragonfly Society of the Americas in 1996.

[All species seen at Turtle Pond unless otherwise noted.]

Damselflies

EBONY JEWELWING *(Calopteryx maculata)*
Rare. Seen several times at the Loch.

AMBER-WINGED SPREADWING *(Lestes eurinus)*
A few were seen in 1998.

SLENDER SPREADWING *(Lestes rectangularis)*
One was identified at the Loch in July 1998, by Ed Lam.

VARIABLE DANCER *(Argia fumipennis)*
Seen once in July 1994 at the Loch.

AZURE BLUET *(Enallagma aspersum)*
Fairly Common.

FAMILIAR BLUET *(Enallagma civile)*
Abundant.

SKIMMING BLUET *(Enallagma geminatum)*
Fairly Common.
ORANGE BLUET *(Enallagma signatum)*
Common.
CITRINE FORKTAIL *(Ischnura hastata)*
Rare. Seen first in May 1998.
FRAGILE FORKTAIL *(Ischnura posita)*
Uncommon.
FURTIVE FORKTAIL *(Ischnura prognata)*
Seen once, September 1996, at Strawberry Fields. There is only one other record for this species in New York State.
RAMBUR'S FORKTAIL *(Ischnura ramburii)*
Seen once, August 1998.
EASTERN FORKTAIL *(Ischnura verticalis)*
Abundant. This is the commonest damselfly in Central Park.
SPHAGNUM SPRITE *(Nehalennia gracilis)*
Seen once, June 1987, Shakespeare Garden.

Dragonflies

SPATTERDOCK DARNER *(Aeschna mutata)*
Seen once, 1998.
BLUE DARNER, unidentified species *(Aeshna sp.)*
A few sightings.
GREEN DARNER *(Anax junius)*
Abundant. This species has the longest flight period of any dragonfly in the park, occurring from late March to mid-November.
COMET DARNER *(Anax longipes)*
Rare.
SWAMP DARNER *(Epiaeschna heros)*
Rare.
UNICORN CLUBTAIL *(Arigomphus villosipes)*
Seen once, June 1987.
PRINCE BASKETTAIL *(Epitheca princeps)*
Seen twice: July 1994 and June 1995.
CALICO PENNANT *(Celithemis elisa)*
Common.

HALLOWEEN PENNANT *(Celithemis eponina)*
Uncommon.

BANDED PENNANT *(Celithemis fasciata)*
Seen once, July 1998, by Janet Wooten.

MARTHA'S PENNANT *(Celithemis martha)*
First sightings in the park, July 1998.

EASTERN PONDHAWK *(Erythemis simplicicollis)*
Fairly common.

SEASIDE DRAGONLET *(Erythrodiplax berenice)*
Several sightings in summer 1998.

DOT-TAILED WHITEFACE *(Leucorrhinia intacta)*
Rare.

SPANGLED SKIMMER *(Libellula cyanea)*
A few seen in 1992, 1998.

SLATY SKIMMER *(Libellula incesta)*
Uncommon.

CHALK-FRONTED CORPORAL *(Libellula julia)*
One seen in 1984, a few in 1998.

WIDOW SKIMMER *(Libellula luctuosa)*
Uncommon.

COMMON WHITETAIL *(Libellula lydia)*
Common.

NEEDHAM'S SKIMMER *(Libellula needhami)*
Uncommon. One sighting in 1987, 1989. Several sightings in 1998.

TWELVE-SPOTTED SKIMMER *(Libellula pulchella)*
Common.

FOUR-SPOTTED SKIMMER *(Libellula quadrimaculata)*
Rare.

PAINTED SKIMMER *(Libellula semifasciata)*
Uncommon.

GREAT BLUE SKIMMER *(Libellula vibrans)*
Uncommon, seen at the Loch and the Gill.

BLUE DASHER *(Pachydiplax longipennis)*
Abundant. The commonest dragonfly in Central Park.

WANDERING GLIDER *(Pantala flavescens)*
Uncommon, usually seen over meadows.

SPOT-WINGED GLIDER *(Pantala hymenaea)*
Common, often seen over meadows.

EASTERN AMBERWING *(Perithemis tenera)*
Very common.
VARIEGATED MEADOWHAWK *(Sympetrum corruptum)*
A few were seen in 1987.
WHITE-FACED MEADOWHAWK *(Sympetrum obtrusum)*
Two sightings in 1989.
RUBY MEADOWHAWK *(Sympetrum rubicundulum)*
Common.
BAND-WINGED MEADOWHAWK *(Sympetrum semicinctum)*
Seen once in 1990, once in 1998.
YELLOW-LEGGED MEADOWHAWK *(Sympetrum vicinum)*
Common. This is the last species to be seen in the fall. May occur until the first frost.
STRIPED SADDLEBAGS *(Tramea calverti)*
The second time this species was sighted in the Northeast occurred in Central Park in 1992. Seen once in 1998.
CAROLINA SADDLEBAGS *(Tramea carolina)*
Uncommon.
BLACK SADDLEBAGS *(Tramea lacerata)*
Common.

Migrating Hawks Over Central Park

Sharon Freedman

Fall is a time for watching migrating hawks. Central Park proved to be an excellent place to run an organized hawkwatch. Our season is from August 15 to December 15. During this time data has been collected on sixteen species of hawks.

While we have the famous red-tailed hawks nesting on Fifth Avenue, there are other species of hawks that nest near Central Park: American kestrels in a few buildings around the park, and peregrine falcons in some structures a mile or two away. They all spend some time in Central Park. In addition to the local hawks, other red-tailed hawks as well as a few sharp-shinned hawks regularly spend the fall and winter here. Cooper's hawk, northern goshawk, and red-shouldered hawk have also made long-term appearances in Central Park and sometimes spend the winter. These hawks are not counted, but do make hawkwatching here that much more exciting.

The following is an account of the species that were seen during our four seasons of watching hawks over Central Park.

BLACK VULTURE *(Cordgyps atratus)*

This primarily southern bird is our rarest regular sighting. It has slowly been expanding its breeding range northward. There has been no set period to observe them. The season record was three.

TURKEY VULTURE *(Cathartes aura)*

This species can be seen throughout the four-month season. Its movement is very light and sporadic in August. The turkey vulture's flight starts to pick up by mid-September and peaks in mid-October. The highest one-day flight, the peak flight, occurred on October 17th. The number of turkey vultures has been increasing over Central Park. The season high was 323.

OSPREY *(Pandion haliaetus)*

Also called a Fish Hawk. Osprey is one of the earliest arrivals and can be spotted from the first day of the hawkwatch season. Osprey can be seen practically every day in September and fairly regularly through mid-October. Their flight starts to slow down after that and by early November it is sporadic. The osprey's migration is over by the end of November. The peak flight was seen on September 19th. The numbers of osprey passing over Central Park has been on the increase. The season record was 851.

BALD EAGLE *(Haliaeetus leucocephalus)*

Mid- to late September is the best time to see this eagle; however, it can be seen almost anytime during the hawkwatch season. It has been one of the highlights of watching hawks over Central Park. There are five age-related plumages of the bald eagle. Any plumage can be seen throughout the four-month period. The highest one-day total of bald eagles of sixteen occurred on September 18th. This level of flight is an exception. In our fourth year there were two days in which nine bald eagles flew over Central Park—September 20th and 21st. The numbers have been variable. The total of sixty-two was our season record.

NORTHERN HARRIER *(Circus cyaneus)*

This beautiful hawk is seen primarily in September and October. Mid-September to mid-October is the best time to view northern harriers. Migration over Central Park is variable and low. The season high was seventy-three.

SHARP-SHINNED HAWK *(Accipiter striatus)*

This little hawk starts its flight in late August and can be seen almost to the end of the season in mid-December. Over Central Park their numbers have been increasing slightly. Sharp-shins can be seen almost every day from approximately September 10th to November 10th. A peak flight of eighty-three occurred on September 24th; however, one-day flights in the fifty- to seventy-plus range occur from mid-September to mid-October. Flights of ten or more are possible through mid-November. The season record was 967.

COOPER'S HAWK *(Accipiter cooperii)*

The first few sightings of this accipiter (a long-tailed woodland hawk) occur in the beginning of September and hit their peak by mid-October. A one-day record of twenty-nine was seen on October 17th. Cooper's hawks can be seen almost until the end of the season; however, they will be in decreasing numbers. The total number of Cooper's hawks has been increasing slightly each year. The season record was 195.

NORTHERN GOSHAWK *(Accipiter gentilis)*

The largest and rarest of the accipiters aren't seen until mid- or late-October. Their numbers are fairly consistent over Central Park. The season record was nine.

RED-SHOULDERED HAWK *(Buteo lineatus)*

This distinctively marked buteo (a broad-winged, short-tailed hawk) isn't usually seen until approximately October 10th, but it can be seen almost until the end of the hawkwatch season on December 15th. A peak flight of twenty-four occurred on November 13th. One-day flights of up to ten are possible until late November. After that time sightings are few and far between. The number of red-shouldered hawks has remained fairly even. The season high was 228.

BROAD-WINGED HAWK *(Buteo platypterus)*

This species provides some of the most exciting viewing of this hawkwatch. They are our earliest buteos and make a sporadic appearance from the beginning of the season. When they really get under way, broad-wings fly in large flocks called "kettles." The one-day high of 5,606 took place on September 19th. This is the one species that seems, under normal circumstances, to have a very precise peak flight date. Sometimes, as in our fourth year, these hawks arrive in very small numbers prior to the nineteenth and then explode in a two-day rush. (On September 20th there were 4,611.) In the second year there was a slow build-up starting on September 10th with an explosion on the nineteenth and then an immediate drop-off in numbers. (From the tenth to the eighteenth numbers ran from 136 to 1,150; on September 20th there were 13.) After the nineteenth, broad-winged hawks make a sporadic appearance until mid-October. The season record was 10,605.

RED-TAILED HAWK *(Buteo jamaicensis)*

This species makes a slow start in mid- to late-September. They trickle in until mid-October when their numbers start to climb. They can be seen almost every day from the beginning of November until the end of the hawkwatch. The peak viewing time for seeing red-tailed hawks migrating over Central Park is from approximately November 10th to

November 25th. The peak flight of 155 occurred on November 24th. The number of red-tailed hawks seen over Central Park has been on the increase. The season record of 1,360 was an exceptionally high number.

ROUGH-LEGGED HAWK *(Buteo lagopus)*

This late-autumn hawk is one of our rarest. It can be observed from early November to the middle of December. The season record was 5.

GOLDEN EAGLE *(Aquile chrysaetos)*

This magnificent eagle was one of the big surprises and controversies of City Hawk Watch. It was not known that golden eagles flew over Central Park until the first one was observed during the first year of the watch. The season record was fourteen.

AMERICAN KESTREL *(Falco sparverius)*

The smallest of the falcons can be seen, in low numbers, from the beginning of the hawkwatch season. The American kestrel can be seen almost every day from the beginning of September until around October 18th. The peak flight of seventy-two occurred on September 18th. The period around the eighteenth seems to be a good time to see American kestrels over Central Park. The seasonal totals have remained fairly constant. The season record was 495.

MERLIN *(Falco columbarius)*

This falcon usually makes its first appearance by late August. The peak flight of ten occurred on September 18th. There was another one-day high of eight which occurred on October 14th. The greatest possibility of seeing a Merlin here would be between September 15th and October 15th. Merlin numbers are low but fairly constant. The season record was fifty-eight.

PEREGRINE FALCON *(Falco peregrinus)*

This large falcon can be seen but not with much regularity. Its numbers are fairly low. A peak flight of six has occurred on three separate dates: September 7th, October 1st, and October 8th. Their seasonal numbers have been fairly consistent. The season record was forty-five.

Schedule for Viewing Hawks Over Central Park			
Species	Earliest Sighting	Latest Sighting	Peak Viewing Period
Black Vulture	9/19	12/15	—
Turkey Vulture	8/15	12/10	9/25 to 10/10
Osprey	8/15	12/6	9/1 to 10/15
Bald Eagle	8/25	12/13	9/18 to 25, 10/5 to 12
Northern Harrier	8/18	12/14	9/15 to 10/20
Sharp-shinned Hawk	8/23	12/13	9/10 to 11/10
Cooper's Hawk	8/23	12/8	9/18 to 10/10
Northern Goshawk	10/15	12/14	—
Red-shouldered Hawk	9/20	12/14	11/10 to 11/20
Broad-winged Hawk	8/15	10/22	9/10 to 9/25, 9/19 peak
Red-tailed Hawk	8/29	12/15	11/10 to 11/25
Rough-legged Hawk	11/4	12/13	—
Golden Eagle	9/25	12/12	—
American Kestrel	8/15	11/12	9/1 to 10/18
Merlin	8/22	11/24	9/15 to 10/15
Peregrine Falcon	8/27	12/13	9/15 to 10/10

A Taste or Two Along the Way

Norma Collin and Charles Kennedy

There are hundreds of edible plants in Central Park. This is simply a list by two people who love the park, who visit it every day the year round and take a taste or two along the way. We do have rules, however. Taste only what you know is absolutely safe (and out of reach of a leg-lifting dog). Take only small amounts. Do not disturb the park in any way to get the taste.

January

FIELD GARLIC: An occasional clump relieves the forager's winter doldrums. The long hollow leaves, sometimes poking out of snow, taste like scallions.

WILD LEEK (or ramp): Our best wild onion. It's strong!

February–March

GARLIC MUSTARD: Our favorite. The young, heart-shaped leaves are delicious raw or slightly steamed.

WINTERCRESS: A little bitter unless cooked.

COMMON PLANTAIN: Cook the leaves while very young, otherwise they're tough and stringy. Can be used as a medicinal brew for stomach troubles.

DANDELIONS: Everywhere. The leaves are great in salads but only before the flower appears. Lots of vitamin A. You can also sauté them, very briefly. Flowers can be dipped in batter and fried. The petals alone make the wine.

April–May

JAPANESE KNOTWEED: The park would like to get rid of this invasive plant, so eat all you want! Usually emerges at beginning of April. That's when the hollow, bamboo-like stalk is tender and tasty when raw. But eat them only when less than a foot tall and peel the skin. Has a nice lemony taste.

COMMON BLUE VIOLETS: Blossoms by end of April.

ELM SEEDS: Nutty and crunchy. Best when light green. Taste differs from tree to tree.

REDBUD: Its small pink flowers one of the best edibles in the park. Redbud tree at the Point has particularly tasty flowers. Try violets, elm seeds, and redbud blossoms for a delicate tiny salad. Wisteria blossoms may be added to this salad.

June

MULBERRIES: Subtle and superb, or just bland? We can't agree—it depends on the tree. The mulberries around Bank Rock Bridge are proof of how good a mulberry can be.

JUNEBERRIES (also called serviceberries or shadberries): They're ripe at June's end, but once we found our first in early July, on the very day the cicadas began to sing. Cedar Hill has them.

BURDOCK: Pull straight up. The taproot is what you want. Peel and cook it. Nutritionally valuable.

WOOD SORREL and SHEEP SORREL: They prove that sour is good, especially in soup and salad.

WINEBERRIES: The park's secret treasure—some say the best berry of all. Sharp competition with birds. Look northwest of Balcony Bridge and southeast of Bow Bridge.

PEPPERGRASS: Pungent and, of course, peppery. When the fireflies light up in mid-July, look for them around the peppergrass.

KOUSA DOGWOOD: Several trees in the Ramble. The bigger the fruit, the better the taste, which is close to mango.

PINEAPPLE WEED: Looks like chamomile, smells like pineapple. We don't really eat this, but its fragrance is as good as a snack.

July

WILD BLACK CHERRIES: Plentiful by the end of the month. Tart to sweet, depends on the tree. Some of the sweetest on a tree around the bend from the Indian Cave, where the woodchuck gorges on celandine in the spring.

August

RASPBERRIES and BLACKBERRIES: They make sneak appearances all through the summer.

September

CORNELIAN CHERRY: Dangerously sour until soft and dark red. Good snack, great jam!

HAWTHORN: Good ones at the top of the Gill. The haws can have an apple-like taste and texture. They're excellent as a jam or jelly. No need for pectin—they've got plenty.

CHAMOMILE, PEPPERMINT, and BERGAMOTS: Blooming now and are ready for tea.

SMARTWEED: Another unloved weed—pick all you want. The tiny bulblets just under the flower are a pretty good, nutty nibble. This is a survival food.

October

AUTUMN OLIVE: Good ones are sweet and tangy. More often they're astringent. Find them at the Point where the warblers love them too.

PERSIMMONS: Several trees northeast of the Evodia Field. Late October is the earliest to sample persimmons and after a frost is the best. Otherwise, painful mouth-puckering effect. To avoid pucker, avoid the tiniest bit of skin. Persimmons are loved by only one of us, but that one loves them well.

HACKBERRIES: They have a small berry—it's a drupe, with sweet, dry pulp surrounding a single pit inside. They have a date-like taste.

SHAGBARK HICKORY: When ripe, the husk turns brown and falls off in four equal parts, revealing a smooth, delicious, but hard to extract nut. But it doesn't matter—the squirrels grab most of the good ones anyhow. They understand.

BLACK WALNUTS: Their location is a secret, for the wood is worth a fortune. The nut tastes like a fortune, but you need a sledgehammer to crack one.

November–December

GINGKO: Follow your nose to find the fallen fruit. It's the outer flesh that smells terrible—the thin-shelled nut inside is tasty when roasted. Wear gloves to gather—the flesh can give you a rash.

CRAB APPLES: Some are the size of marbles or cherries, some almost the size of little apples. Some are mealy; some are firm and tart. They ripen earlier, but our favorites are picked from the tree in November, when they're soft and brown and mushy and taste like fine applesauce.

At Various Times, Spring or Late Fall

SULPHUR SHELF (or chicken mushrooms): The only Central Park fungus we recommend. You can't confuse it with anything else. Grows on dead or injured deciduous trees, logs, or stumps. The flesh is bright orange and yellow, and sometimes there are pounds of it.

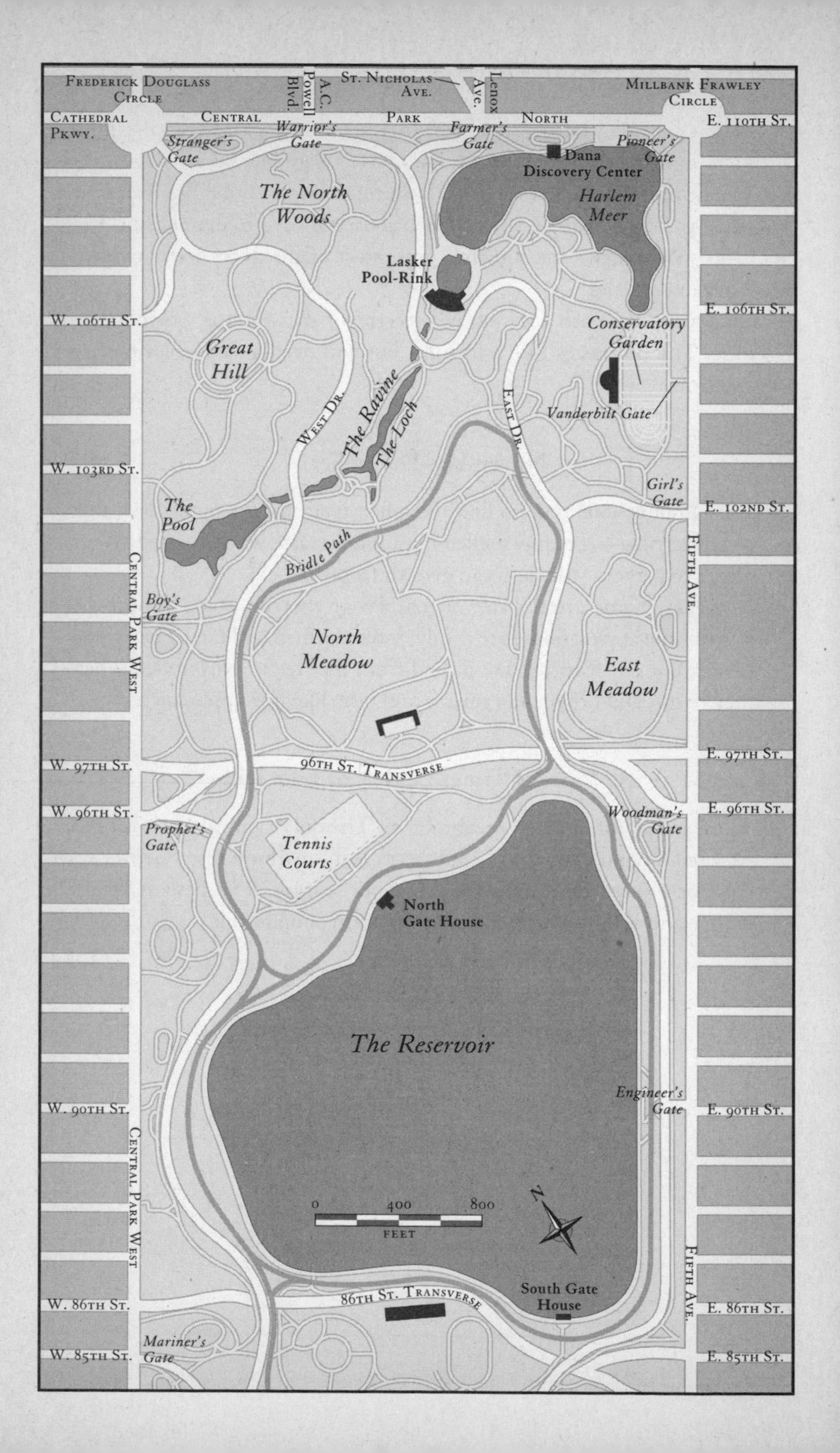

Frederick Douglass Circle
A.C. Powell Blvd.
St. Nicholas Ave.
Lenox Ave.
Millbank Frawley Circle
Cathedral Pkwy.
Central Park North
E. 110th St.
Warrior's Gate
Farmer's Gate
Stranger's Gate
Pioneer's Gate
Dana Discovery Center
Harlem Meer
The North Woods
Lasker Pool-Rink
W. 106th St.
E. 106th St.
Conservatory Garden
Great Hill
The Ravine
The Loch
West Dr.
East Dr.
Vanderbilt Gate
W. 103rd St.
Girl's Gate
The Pool
E. 102nd St.
Bridle Path
Fifth Ave.
Central Park West
Boy's Gate
North Meadow
East Meadow
W. 97th St.
96th St. Transverse
E. 97th St.
W. 96th St.
Woodman's Gate
E. 96th St.
Prophet's Gate
Tennis Courts
North Gate House
The Reservoir
Engineer's Gate
W. 90th St.
E. 90th St.
0
400
800
Feet
N
South Gate House
W. 86th St.
86th St. Transverse
E. 86th St.
Mariner's Gate
W. 85th St.
E. 85th St.

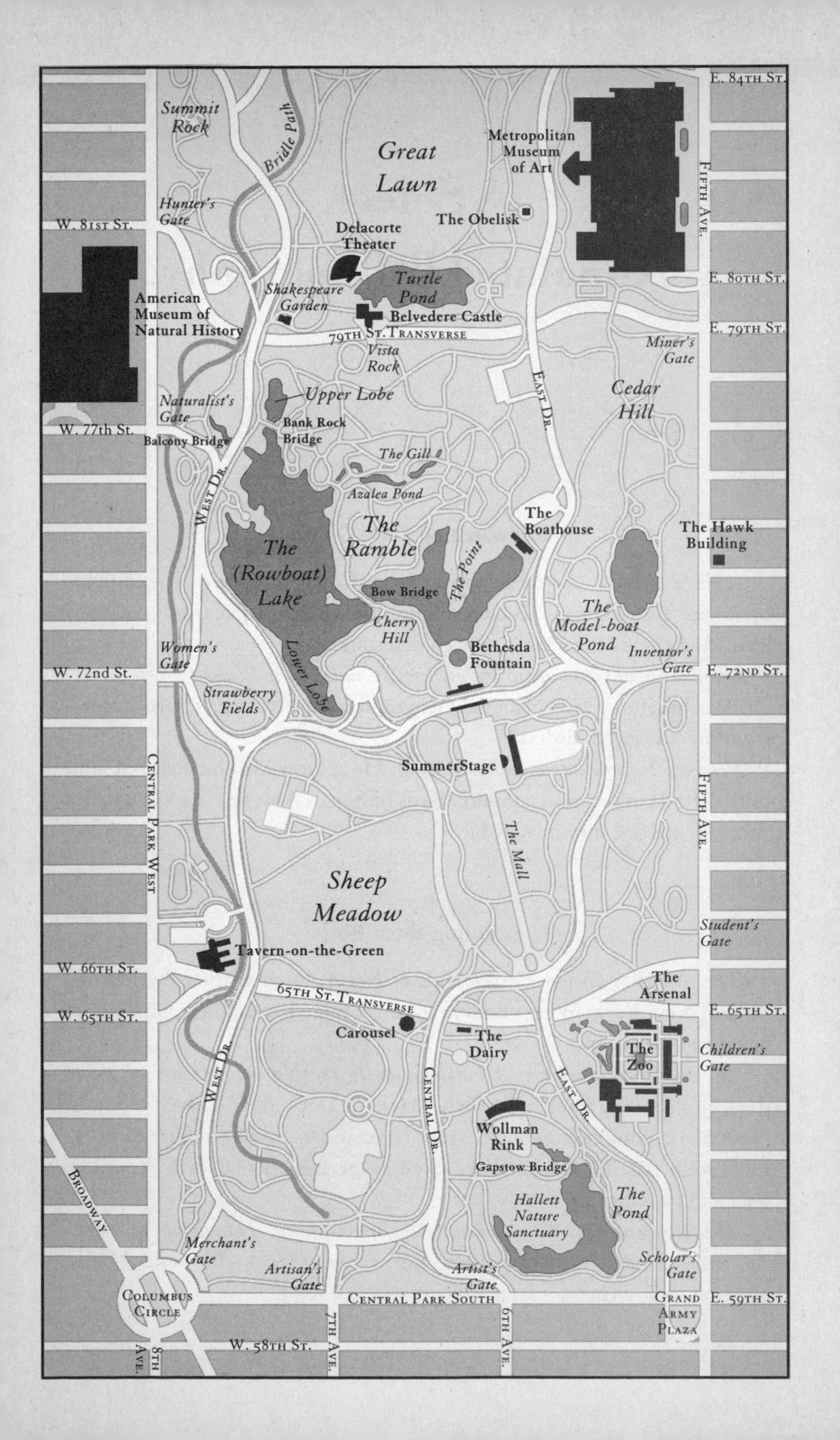

Summit Rock
Bridle Path
Great Lawn
Metropolitan Museum of Art
E. 84th St.
Fifth Ave.
Hunter's Gate
W. 81st St.
Delacorte Theater
The Obelisk
Turtle Pond
E. 80th St.
American Museum of Natural History
Shakespeare Garden
Belvedere Castle
E. 79th St.
79th St. Transverse
Miner's Gate
Vista Rock
East Dr.
Cedar Hill
Upper Lobe
Naturalist's Gate
Bank Rock Bridge
W. 77th St.
Balcony Bridge
The Gill
West Dr.
Azalea Pond
The Boathouse
The Ramble
The Hawk Building
The (Rowboat) Lake
The Point
Bow Bridge
The Model-boat Pond
Cherry Hill
Women's Gate
Bethesda Fountain
Inventor's Gate
W. 72nd St.
E. 72nd St.
Lower Lobe
Strawberry Fields
SummerStage
Central Park West
Fifth Ave.
The Mall
Sheep Meadow
Student's Gate
Tavern-on-the-Green
W. 66th St.
The Arsenal
65th St. Transverse
E. 65th St.
W. 65th St.
Carousel
The Dairy
The Zoo
Children's Gate
Central Dr.
East Dr.
West Dr.
Wollman Rink
Gapstow Bridge
Broadway
Hallett Nature Sanctuary
The Pond
Merchant's Gate
Artisan's Gate
Artist's Gate
Scholar's Gate
Columbus Circle
Central Park South
Grand Army Plaza
E. 59th St.
W. 58th St.
8th Ave.
7th Ave.
6th Ave.

A Select, Annotated Bibliography

Birdwatching Field Guides

A Field Guide to the Birds, 4th Edition (Eastern and Western editions) by Roger Tory Peterson. Boston: Houghton Mifflin, 1980. (Referred to by birdwatchers as "Peterson"—the golden standard.)

The National Geographic Society Field Guide to the Birds of North America. Washington, D.C.: National Geographic Society, 1993. ("Geo": useful for more advanced birders.)

Peterson First Guide to Birds. Boston: Houghton Mifflin, 1986. (A simplified guide to the most common birds of North America, a good start for children but fine for any new birder.)

Learning about Birds

The Audubon Society Encyclopedia of North American Birds by John K. Terres. New York: Wings Books, 1991. (A fine all-purpose compendium of bird information.)

The Birder's Handbook: A Field Guide to the Natural History of North American Birds by Paul R. Ehrlich, David S. Dobkin, and Darryl Wheye. New York: Simon & Schuster, 1988. (An indispensable reference book, filled with detailed information about every species of North American bird and its habits.)

The Birdwatcher's Companion: An Encyclopedic Handbook of North American Birdlife by Christopher Leahy. New York: Portland House, 1997. (A well-written, informative reference book.)

Hawks in Flight by Pete Dunne, David Sibley, and Clay Sutton. Boston: Houghton Mifflin, 1988. (A new way of identifying hawks at a distance, when field guides don't help.)

How Birds Migrate by Paul Kerlinger. Mechanicsburg, PA: Stackpole Books, 1995. (An excellent overview of a complicated subject; reads well.)

Ornithology, 2nd edition, by Frank B. Gill. New York: W. H. Freeman, 1995. (A scientific text for non-scientists—a useful resource for those who have gone beyond identifying birds.)

Watching Birds, An Introduction to Ornithology by Roger Pasquier. Boston: Houghton Mifflin, 1977. (A well-written, accessible introduction to bird studies by a Central Park birdwatcher.)

Trees and Flowers

A Field Guide to Trees and Shrubs by George Petrides. Boston: Houghton Mifflin, 1972. (Well-organized guide; consulting five different tree guides helps.)

A Field Guide to Wildflowers by Roger Tory Peterson and Margaret McKenny. Boston: Houghton Mifflin, 1968. (A classic, organized by color.)

Identifying and Harvesting Edible and Medicinal Plants by "Wildman" Steve Brill. New York: Hearst Books, 1994. (The author leads foraging tours in many New York City parks, including Central Park.)

Butterflies

Butterflies Through Binoculars: Boston, New York, Washington Region by Jeffrey Glassberg. New York: Oxford University Press, 1993. (A new way to study butterflies, not for beginners and only for East Coast enthusiasts, but may start a much-needed revolution.)

Stars and Planets

The Stars: A New Way to See Them by H. A. Rey. Boston: Houghton Mifflin, 1980. (A great guide for beginners by the author of *Curious George.*)

Thoreau

The Journal of Henry D. Thoreau: In Fourteen Volumes Bound as Two. New York: Dover Publications, 1962. (Thoreau's journal, written between 1837 and 1861, began as a twenty-year-old's book of jottings and ended as the great writer's supreme masterpiece. One could spend a lifetime studying these books, full of detailed observations of the natural world and thoughts about man's place in it.)

More about Central Park

"The Birds of Central Park: An Annotated Checklist by the Birdwatchers of Central Park," compiled and designed by Rebekah Creshkoff and Marie Winn. A publication of the Central Park Conservancy, 1996. (Available free at the Castle, the Dairy, the Dana Center, and other places in the park.)

Bridges of Central Park by Henry Hope Reed, Robert M. McGee, and Esther Mipaas. New York: The Greensward Foundation, 1990. (Every bridge located, explained, and illustrated. Includes a succinct but thorough history of the creation of Central Park.)

"The Elliott Newsletter: Nature Notes from Central Park" by Sarah Elliott, 333 East 34th Street, New York 10016. (A monthly round-up of everyday happenings in the park's nature world, focusing mainly on birds, insects, trees, and flowers. By subscription.)

Nature Walks of Central Park by Dennis Burton. New York: Henry Holt, 1997. (Enjoyable and instructive walking tours of the park, with a special focus on trees and flowers, by the park's woodlands manager.)

Acknowledgments

ONE DAY ALMOST TEN YEARS AGO I ran into Raymond Sokolov, editor of the Leisure and Arts page of the *Wall Street Journal.* "Why don't you write a column about birds for my page," he said and I was off and running on a new professional life. Writing for the *Journal* and for the eye and ear of a polymath like Raymond Sokolov has been a unique pleasure in my career.

What I owe to Norma Collin, Tom Fiore, Sharon Freedman, Charles Kennedy, and Nick Wagerik (in alphabetical order) is evident throughout this book; I hope some of the affection I feel for them has also managed to show through.

Special thanks are due to Dorothy Poole and Anne Shanahan, who appear in the story more briefly than their importance to it merits; so too does David Monk, whose prodigious knowledge of the park's trees was a great resource; Peter Post, Marty Sohmer, and Dick Sichel gave me a useful historic perspective of birding in Central Park.

Jim Lewis, Holly Holden, and Blanche Williamson shared their observations about the Fifth Avenue hawks with me, and we spent many hours of hawk obsession together. Thanks are due to Merrill Higgins for helping to solve the mystery of a hawk's identity, and to Barbara Ascher for letting me use part of her prayer for the hawks; and to Michael O'Gara, astronomer and lieder-singer extraordinary.

I'm indebted to other park friends: Mary Birchard, Mike Bonifanti, Mo and Sylvia Cohen, Bill DeGraphenreid, Joe and Mary Fiore, Tom Flynn, Chris Girards, Evan Kamil, George Muller, Margaret O'Brien, Harold Perloff, the ever present spirit of Joe Richner, Ira Weil, Bob Woods, Janet

Wooten, Elliott Zichlinsky; the Earlybirds: Karen Asakawa, Rhoda Bauch, Shale Brownstein, Cousin Fred (until he moved to Pennsylvania), Patti Formisano, Marianne Girards, Michi Kobi, Jane Koryn, Richard Rabkin, Deborah McMillan, Estelle Symons, Irene Warshauer.

Marianne Cramer, Park Planner for the Central Park Conservancy, provided maps, historical information, encouragement, guidance. Informed by her knowledge and her love for Central Park, the Ramble will one day be revitalized without any loss of its value as a wildlife habitat. Thanks also to Neil Calvanese, now Chief of Operations for the park—my admiration for his horticultural know-how and his love of growing things is boundless. To Dennis Burton, Central Park's Woodlands Manager: Thanks for making the North Woods an inviting place for birdwatchers. Thanks to Regina Alvarez, the zone gardener for the model-boat pond area while much of this story was going on. She won the hawkwatchers' hearts as the hawks won hers. Thanks also to Chris Seita, zone gardener of the Shakespeare Garden, whose sharp eye discovered the chrysalis of a black swallow-tail butterfly on a spiderwort leaf—a needle in an Everest-sized haystack.

To Paul Sweet, research assistant in ornithology at the American Museum of Natural History, and to Clare Flemming, his counterpart in the mammology department—thanks for illuminating the arcane mysteries of bird and mammal physiology, morphology, and several other -ologies. To Cal Snyder of the entomology department, thanks for the four "summer Bug Nights" in the Ramble, where we saw a girlfriend underwing and a greater black-letter dart (and the man in a Dracula costume who suddenly materialized out of the dark of the Ramble and stayed to admire the above-mentioned moths).

In gathering material about the hawk story, I was helped by Fifth Avenue and Central Park West neighbors of the hawks: Mary Tyler Moore, Linda Janklow, Jane Koryn, Pamela Clauss, Stanley Diamond, Dorris Carr, Judith Hernstadt, Joan Schwartz (belated thanks for all the photos of Pale Male and Mom I or II perched on a living room windowsill, many of them sent me over the years).

Thanks are due to the various professionals who generously spent time answering questions: Dean Amadon, Jerome A. Jackson, Paul Kerlinger, Paul Kupchok, James Kushlan, Judith McIntyre, Chris Nadareski, Roger Pasquier, Charles Preston, Steve Quinn, Ward Stone, John K. Terres, Julio de la Torre.

ACKNOWLEDGMENTS

Len Soucy, inspiring wildlife rehabilitator, is at the heart of the Fifth Avenue red-tail story—both female hawk heroines were rescued and returned to Central Park by this remarkable man. I thank him not only for the many interviews he granted, and for the exhilarating afternoon I spent at the Raptor Trust, but for the thousands of birds, from great hawks to small sparrows, that he has nursed back to health and returned to the wild.

To Rebekah Creshkoff, a Central Park Regular and a birdwatching companion I have learned much from: thanks for years of encouragement and help, for meticulously checking the final version, and for sending the *Far Side* cartoon about writers as sheep. To Martha Miller, thanks for the chocolate kisses.

To Barbara Dubitsky, thanks for years of listening to stories. To Sarah Paul and Sarah Kahn, thanks for much good advice and for friendship.

Tom Fiore read the book in manuscript form and corrected numbers of errors. Nick Wagerik read portions of the book and made valuable suggestions. Neither of them is responsible in the least for mistakes that may appear here.

For help and encouragement in early stages of this book that permanently affected its final outcome I am grateful to Bobbie Bristol—this book would not exist without her.

In the later stages of writing, I was helped immeasurably by Eric Bentley, who offered a peaceful room to work in when files and books and papers and the spirit of confusion threatened to engulf me.

Dan Frank at Pantheon is an editor of the kind they say no longer exists, one passionately devoted to good writing and one who has uncanny skills in helping writers overachieve. His taste and intelligence (and blue pencil) are reflected throughout this book.

Others at Pantheon whose help I enjoyed: Claudine O'Hearn, Kristen Bearse, Kathy DiGrado, Altie Karper, and very special thanks to Grace McVeigh for help beyond the call of duty.

Thanks to Anne Malcolm for the magical endpapers, and for being a true Regular in her heart. To Mike Miller, thanks for brilliant suggestions, and for giving me the deepest understanding of what the hawk Moms felt when their kids left the nest.

Finally, for advice both literary and cinematic (keep the story moving!) and for encouragement beyond description, I owe thanks and love to Allan Miller.

Index